THE
WEATHER MATRIX
AND
HUMAN BEHAVIOR

THE WEATHER MATRIX AND HUMAN BEHAVIOR

Michael A. Persinger

PRAEGER

PRAEGER SPECIAL STUDIES • PRAEGER SCIENTIFIC

Library of Congress Cataloging in Publication Data

Persinger, Michael A

    The weather matrix and human behavior.

    Bibliography: p.
    Includes index.
    1. Psychology, Physiological. 2. Weather--Mental
and physiological effects. 3. Man--Influence of
climate. 4. Human behavior.   I. Title.
QP360.P46   612'.0144   80-18422
ISBN 0-03-057731-4

Published in 1980 by Praeger Publishers
CBS Educational and Professional Publishing
A Division of CBS, Inc.
521 Fifth Avenue, New York, New York 10017 U.S.A.

© 1980 by Praeger Publishers

0123456789   145   987654321

Printed in the United States of America

PREFACE

The Weather Matrix and Human Behavior is a behavioral and
biological approach to the complex problem of classical biometeorology. Significant emphasis is also placed upon control systems
theory, a concept compatible with behavioral principles. With these
techniques, problems and possibilities for the young environmental
scientist are developed.

The book is written especially for the undergraduate or graduate
student who appreciates the necessity of the "integrator scientist:"
the individual with fundamental comprehension of the basic principles
of physics, chemistry, geology, meteorology, physiology, psychology,
and statistics. This individual will be the key component of science
as overspecialization fragments knowledge into smaller and smaller
increments.

Although the integrator scientist cannot hope to memorize the
details of the specialist, a general understanding of these basic
principles and of the techniques by which they can be applied is by
far a more critical commodity. The accelerating availability of
computers and data files will soon allow the integrator scientist to
approach complex environmental and interdisciplinary problems
with prowess equal to that of the specialist.

The weather matrix, characterized by many meteorological,
geophysical, astronomical, and idiosyncratic factors, is an excellent
field for stimulus complexity. The human locus of behaviors, another
matrix of varying, intercorrelated events, is the important point
immersed within the weather matrix. Used wisely, the study of the
relationship between the weather matrix and human behavior can be
a central theme around which the principles of many sciences can
be integrated and understood.

I thank Angela Corradini, Patricia McMahon Lepage, Jackie
Beaudin Stanley, and Lynn Suess for their technical assistance during
the preparation of the manuscript. Special acknowledgement is given
to Claudette Larcher for her help and time. My gratitude to George
P. Zimmar and to Praeger Scientific for appreciating the potential
importance of human behavior and the weather matrix.

CONTENTS

Page

PREFACE   v

Chapter

1   SOURCES AND VARIATIONS OF WEATHER
    STIMULI   1

    The Air Mass: Unit of Weather   2
        Fronts: Interfaces of Air Masses   3
        Components of the Weather Matrix   3
    Candidate Stimuli of the Weather Matrix   6
        Temperature   6
        Humidity   7
        Barometric Pressure   8
        Wind   9
        Ions   10
        Electric Fields   11
        Electromagnetic Fields   12
    Indirect Consequences of the Weather Matrix   13
        Soil Gases   13
        Organic Materials   13
        Man-made Pollutants   14
        Geophysical Effects   14
    Astrogeophysical Correlates of the Weather Matrix   15
        Geomagnetic Storms   15
        Lunar Phase   15
        Infrasound   16
        Solar Disturbances   16

2   CONCEPTUAL SITES OF ACTION WITHIN
    THE HUMAN LOCUS   17

    High Gradient Sites of Weather Stimulation   18
        The Central Nervous System   18
        The Hypothalamus: Emotional Integrator   21
        The Autonomic Nervous System   23
        The Cardiovascular System   26

The Blood   26
The Bones   27
Particular Organs   28
Low Gradient Systems   28
The Water Molecule   29
The Cell Membrane   30
The Ground Substance   31
Loose Connective Tissue   32
Body Mechanical Vibrations   33
Steady State or Direct Current (DC) Electric Fields   33
Time-varying Electromagnetic Fields   34

3   THE HUMAN ORGANISM AS A COMPLEX
     ENSEMBLE OF FEEDBACK SYSTEMS:
     EXPECTED PROPERTIES   35

Simple Feedback Systems   35
A Mechanical Example   37
Biological Examples   37
Behavioral-Psychological Examples   38
Intrinsic Properties of Homeostats   40
Failures in Biological Systems   42
Relevance to Weather   44
Basic Signal Shapes and Patterns of Response   44
Step Signals   44
Ramp Signals   46
Impulse Signals   47
A Note on Time Intervals   48
A Special Class of Signals   48
Factors Controlling the Intensity of Weather Responses   50
The Law of Initial Values   50
Compensatory Overshoots   53
Response Exhaustion: Single Unit   54
Response Exhaustion: Populations   55
Periodic Variations in Baseline   56
Ultrashort Periods   57
Short Periods   57
The Circadian Clocks   57
Long Periods   58
Seasonal Clocks   59
Annual Clocks   59
Infralong Periods   59
General Concepts of Weather Stimuli and Systems Theory   59

4   LANGUAGE AND LEARNING AND THE
      HUMAN LOCUS   62

      Learning and Conditioning   62
            Respondent Conditioning   64
            Operant Procedures   69
            Superstitious Conditioning   70
            Unconscious Conditioning   72
      Application to Biometeorology   76
            What are the Unconditioned Stimuli in Weather?   79
            A Radical Conditioning Option   80
            Superstitious Conditioning   81
            The Time Factor   82
      Language and Thinking and the Body Locus   83
            Language as Measurements of Events   84
            Language as a Label of Organismic Events   85
            Types of Descriptors   88
            Problems of Discrimination   89
            Words as Determiners of Behavior   89
      The Placebo or Expectancy Effect   91
            Expectancy and Private Behaviors   91
            Egocentricity and Explanation   93
            Placebos and Personality   93
            Diseases Treated by Expectancy   95
            Application to Biometeorology   96

5   TYPICAL WEATHER CORRELATES AND
      PROBLEMS OF CLEAR DEMONSTRATION   98

      General Patterns of Human Weather Responses   99
            Typical Behavioral Complaints   99
            Typical Discrete Behavioral Changes   100
            Biological Responses   100
            Relationship to Specific Weather Changes   100
            Magnitude of Weather Effects   101
      General Methods of Demonstration   102
            The Correlational Method   102
            The Pearson Product-Moment Correlation   104
            Practical Significance   105
            Correlation is not Causation   109
            The Third Factor   110
            The Problems of Causality   110

Factors Limiting Correlational Studies   112
        The Confounding Factor   112
        Single-Factor Correlation   113
        Seasonal Effects   114
Suggested Procedures for Correlational Studies   114
        Use the Weather Matrix   114
        Use Absolute and Rate of Change Measures   115
        Correlate the Weather Matrix   115
        Use More Than One Behavioral Measure   115
        Lag Correlations   115
        Watch for Intrinsic Variability   116
        Watch for Curvilinear Relationships   116
        Watch for Weekend Effects   118
        Replicate   118
The Weather-Sensitive Person   118
        The Obscuring Average   120
        Studies Based upon Complaints   120
        Limits of the Term   121
        Theoretical Bases of Weather Sensitivity   121
        Percentage of Weather-Sensitive People   123
Phases of the Weather Matrix   123
        Examples   124
        Importance of Weather Phases   126
        Advantages and Disadvantages   127

6 TEMPERATURE, HUMIDITY, AND RELATED
   CONDITIONS   129

Temperature: The Perspective   129
        Measurement of Temperature   130
        Biological Significance   131
        Simple Heat Physics and Physiology   133
        Temperature Plus Time Problems   134
Seasonal Temperature   139
        Human Behavior and Seasonal Temperature
                Variations   139
        Heart Failure and Stroke   144
        The Data   146
        Time Lagging   149
        Confounding Factors in Seasonal Temperature
                Effects   149
Protracted Periods of Extreme Temperatures   151
        The Heat Wave   152

Physical Patterns of Heat Waves and Mortality   152
Types of Death Categories Associated with
    Excessive Heat   154
The Magnitude of the Effect   155
Cold and Warm Fronts   156
    Theoretical Consequences   156
    Organismic Responses   157
    Temperature, the Common Cold, and Influenza   157
    Mortality   161
    Eclampsia   161
    Critical Duration of Exposure   162
Humidity: Water Vapor in the Air   163
    Measures of Humidity   163
    Relevance of Humidity   164
    Humidity and Respiratory Complications   166
    Comfort   168
Some Special Conditions   169
    Rain   169
    Cloudy Days   170
    Brightness   171
Some General Behavior Neurobiological Effects of
Temperature   172
    The Thermal Mesh of the Body Locus   173
    The General Theory   174
    The Limbic Connection   176
    The Limbic–Prefrontal Cortex Interface   179
Nonthermal Stimuli   180

7   MECHANICAL STIMULI OF THE WEATHER
     MATRIX: BAROMETRIC PRESSURE, WIND,
     AND INFRASOUND   182

Barometric Pressure   182
    Temporal Variations   184
    Sample Correlational Studies   186
    Critique of Correlational Studies   187
    The Hollander Experiments   189
    A General Mechanism   191
Wind   193
    Characteristics   194
    Organismic Relevance   195
    Correlational and Experimental Studies   197
Special Winds   197

Infrasound   199
    Characteristics of Normal Sound   199
    Characteristics of Infrasound   200
    Sources of Infrasound   201
    Man-made Sources of Infrasound   202
    Biobehavioral Relevance   202
    Correlational Studies   203
    Correlational Criticisms   203
    Experimental Studies   204
    Problems with Experimental Studies   204
    Possible Mechanism of Detection   205

8   ELECTROMAGNETIC STIMULI OF THE WEATHER
    MATRIX: ELECTRIC FIELDS AND ATMOSPHERIC
    IONS   207

Electrostatic Fields   207
    Variations in Electric Field Values   208
    Weather-related Variations   208
    Biometeorological Relevance   209
    Correlational Data   210
    Experimental Studies   210
    Steady Potentials of the Human Body   211
    Induced Atmospheric Electric Field Currents 212
Small Air Ions: Particulate Electric Fields   213
    Types of Ions   214
    Concentrations and Sources   214
    Biometeorological Relevance   215
    Correlational Studies   216
    Experimental Studies   217
    The Everpresent Expectancy Effect   219
    The Krueger Studies   221
    The Serotonin Hypothesis   222

9   ELECTROMAGNETIC STIMULI OF THE WEATHER
    MATRIX: MAGNETIC FIELDS AND EXTREMELY
    LOW FREQUENCY ELECTROMAGNETIC FIELDS   227

Geomagnetic Fluctuations   227
    The Magnetosphere   228
    Geomagnetic Storms   229
    Periodicities in Geomagnetic Activity   232
    Geomagnetic Activity Correlates with Weather 233

Biometeorological Relevance   235
Correlational Studies   235
Criticisms of Human Correlational Studies   236
Animal Correlational Studies   237
Extremely Low Frequency Electromagnetic Fields   240
Characteristics   240
Biometeorological Relevance   243
Correlational Studies   244
Experimental Studies with Magnetic Fields   245
Human Experiments   247
Nonhuman Animal Studies   251
Magnetism Mania in Theory   257

10   SOLAR, LUNAR, AND POTENTIAL PLANETARY
      CORRELATIONS WITH HUMANS AND THE
      WEATHER MATRIX   261

The Sun   262
Monochromatic Sunspot-related Phenomena   264
The Sunspot Cycle   264
Flares: Short-term Solar Flux Variations   265
The Solar Constant   266
Principles of Solar-Terrestrial Interaction   267
Solar-Weather Correlations   268
Solar Activity and Living Systems on Earth   271
The Ubiquitous 11-Year Cycle   271
Mechanisms of Solar-Weather Interaction   272
The Moon   273
Lunar Periodicities and Phases   273
Biometeorological Relevance   274
Lunar Correlations with Atmospheric Phenomena   275
Potential Mechanisms of Lunar-Earth Weather
  Relationships   276
Lunar/Nonhuman Animal Correlations   277
Lunar Phase Correlates of Human Behavior   278
Theories of Lunar Effects   280
Explanations of Lunar Effects   284
Planetary-Solar-Terrestrial Correlations   285
The Correlations   286
The Nelson Report   287
Direct Planetary Modulation of Terrestrial
  Phenomena   288
Bioplanetary Correlations   289

Chapter   Page

11   PERSPECTIVE AND POSSIBILITIES   291

      General Summary   292
           Temperature   292
           Behavior   294
           Selective Population   295
      Have We Measured the Appropriate Behaviors?   297
      The Requirement for a Response Matrix Analysis   299
      Psychoepidemiology and Weather   300
      The Tenuous and the Ubiquitous   304

REFERENCES   307

INDEX   321

ABOUT THE AUTHOR   329

THE WEATHER MATRIX AND HUMAN BEHAVIOR

1

SOURCES AND VARIATIONS
OF WEATHER STIMULI

Around the earth is a tenuous complex of suspended particles and diffusing gases incessantly penetrated by a vast range of electromagnetic energies. The major gaseous components of this complex, such as nitrogen (78 percent by volume) and oxygen (21 percent), display stable concentrations while other components fluctuate radically due to biological and geophysical reactions. Electromagnetic energies in the form of heat excite the gases and contribute to their massive movements across the earth's surface, while other forms, such as static electric fields, determine the vertical movements of charged particles.

The human species is immersed in this complex. Each of us is constantly engulfed by it. The components press against our skin, invade our lungs, and may even pervade our bodies. A few changes in the atmosphere are easily perceived. We readily observe the clouds and feel the wind. Sometimes we respond to extremes in temperature or humidity. Other changes are less easily perceived, but they can evoke powerful alterations in individual and group behaviors.

Changes in this environmental complex have been given two major descriptors: weather and climate. The two concepts differ more in temporal properties and absolute magnitude than in basic constituents. Weather can be described as atmospheric changes over a few days, while climate involves months or even years. Upon the powerful baseline of climate, weather appears as a short-termed shallow ripple. The effects of weather upon humans are as complex as its sources.

THE AIR MASS: UNIT OF WEATHER

The constituents of the atmosphere are not confined within the same space. Energy from the sun pounds incessantly upon these constituents to induce macrocosmic motion over large areas. Since the amounts of energy impinging upon different areas of the earth's surface are not equal, atmospheric motions are induced. Areas of different energy levels, as measured by temperature or barometric pressure, move in relation to each other according to well-established principles. These motions are influenced as well by the vortical effects from the earth's rotation and by the variable complexion of the terrestrial surface.

The earth's surface or topography also influences the overlying air by contributing to its physical and chemical characteristics. Areas where this occurs are called source areas, which represent the simplistic dichotomy of temperature and matter. The two extremes of temperature are cold (polar) and hot (tropic), while the two extremes of matter (at least on the earth's surface) are land (continental) and water (maritime). These source areas are usually stable for sufficient periods to allow the overlying air to obtain more or less homogeneous properties.

Partitions of air with similar physical and chemical properties are called air masses. Air masses are usually typed by their temperature, humidity, and barometric properties, which, in turn, traditionally reflect the source area. Consequently, a polar maritime air mass would show cold temperatures with high moisture content. On the other hand, a continental polar air mass would display low temperatures with relatively low moisture (low humidity). Polar sources may be characterized by high barometric pressures, while tropical air masses would show comparatively lower barometric pressures. These general designations are usually divided into more useful subcategories that allow more precise description of the great variety of air masses.

Air masses are large compartments of atmosphere. These regions range from a few hundred kilometers to a few thousand kilometers wide and long but are only a few kilometers thick. Metaphorically, one can picture an air mass as a very wide, extremely thin blob of atmosphere, with fluid characteristics, slowly moving over the earth's variable surface. Sometimes an air mass moves over the oceans, while at other times it moves over the land's bumpy surface. The disparity of energy distribution between these air masses in conjunction with the earth's rotational forces maintains their unending movements. As these vast matrices of atmosphere move over the terrestrial surface biological systems are totally immersed in their properties.

Fronts: Interfaces of Air Masses

Air masses do not usually mix with each other in a technical sense. Instead they remain separated by a boundary or interface that moves with them. The actual width of the boundary or frontal region frequently reflects the movement of the most energetic air mass. If cold air moves toward warm air, a cold front is produced with a slope of about 1:30 or 1:100. In this instance, a cold air mass one kilometer in height, for example, would move from 30 to 100 kilometers beneath a warm air mass. If warm air moved toward a cold air mass, the slope would be about 1:100 to 1:400. In this situation, warm air one kilometer thick would extend over from 100 to 400 kilometers of cold air.

Great exchanges of energy can occur within these bands that separate cold and warm air masses. In general, the greater the disparity in temperature and barometric pressure between two closely adjacent air masses, the more energetic and unstable the exchange. Large scale movements of warm air in relation to relatively cold air are perceived as wind. When the same type of energies are focused within a very small area a few kilometers in diameter, intense but short-termed episodes of turbulence, such as thunderstorms, tornados, and windstorms, can occur. Such loci of instability may occur spontaneously, or so it may seem, along the front, quickly forming and dissipating.

Frontal systems represent more or less sudden changes in weather factors to which animals, including human beings, must adjust. During such periods of adjustment, the living system can display physiological, chemical, endocrine, humoral (blood-related), or even obvious behavioral responses that occur only or with greater intensity during these periods. Following sudden decreases in temperature, mammals display transient increases in general activity, such as ambulation, talking, or agitation. Following sudden decreases in barometric pressure, animals may display lethargy and general decreases in mobility. Once the living system has adjusted to the weather changes after several hours or days of continuous exposure, the same weather factor may not evoke comparable behaviors. Many biometeorological phenomena exist as transient response episodes displayed during adjustment to sudden weather changes.

Components of the Weather Matrix

A given air mass contains a variety of different physical and chemical parameters that can stimulate the human being. They include temperature, relative humidity, barometric pressure, wind speed, air ions, electromagnetic waves, aerosols, and organic materials. In addition, these factors can influence local geophysical conditions to release soil gases, alter the local electric field, or change the relative acidity of the air.

A central problem in biometeorology arises when researchers attempt to isolate the physical or chemical variable most responsible for reported biological or psychological effects. Many of these variables do not change in a simple or isolated manner, but display time variations together. For example, the important triad of temperature, relative humidity, and barometric pressure displays highly interrelated changes. When one variable changes so do the other two. Depending upon season and locality, the direction of the correlations within the triad as well as with other environmental variables can demonstrate very different patterns. Attempts to isolate the important factor become precarious.

This complex of highly intercorrelated but time-varying clusters of atmospheric events can be considered the weather matrix. The term weather matrix is preferred since there are numerous candidates within a given air mass capable of inducing or influencing biobehavioral changes in human beings. Although one could demonstrate specific correlations between a single stimulus and behavior, such as the sudden increases in human aggression associated with increases in relative humidity during summer seasons, many other environmental factors are also changing.

Single correlational analyses are prone to a critical intrinsic flaw. Even when two variables are highly correlated, there may still be a third factor—unspecified or not included in the observations— that is responsible for the apparent relationship between the other two variables. This statement, although a useful adage for the young scientist, is dangerous at the extremes. At the most conservative end of interpretation, one might conclude that all correlations are worthless. Within a more liberal form, one might fall into an untestable mysticism whereby the mysterious third factor is always responsible.

Selection of a single factor from the weather matrix as "the explanation" or "the cause" of human behavior ignores the vast complexity of the issue. Not only are factors in the weather matrix intercorrelated in space; they are also correlated in time. For example, disturbances in the earth's magnetic field frequently occur from two to three days before unstable weather conditions. In turn, significant changes in solar activity can occur from one to three days before the geomagnetic disturbance. What would be the actual control stimulus for any odd or weather-specific behavior associated with these periods? Would it be one of many variables changing on the day of the behavioral alterations? Or would the actual stimulus be some geophysical or solar factor that occurred hours or days before the weather change? Close temporal proximity between a weather change and human behavior does not prove cause and effect. Some biological responses may require days to be elicited.

The complexity and potential error of correlational analyses and experimental manipulation of single elements of the weather matrix become evident when a final factor, cluster effects, is considered. Biometeorologists still do not know how clusters of weather stimuli changing at more or less the same time influence living systems. Do combinations of stimuli within the weather matrix cancel each other's effects? Or does each component of the weather matrix slowly add its effect in an incremental manner to the total impact upon the living system? Taken singly, individual weather components may display no discernible effects upon human beings. Considered as a matrix, weather components could produce cumulative and blatant alterations in the stability of human behavior.

A natural consequence of combining more than two environmental factors is the formation of interactions, potential effects or stimuli that occur only when two or more factors occur together. Interactions are well known in fields such as pharmacology, where they are called synergistic effects. One lethal example is the presentation of chloroform to an animal injected with adrenalin. Adrenalin by itself induces only increases in heart rate, while chloroform presented singly induces sleep. When the two drugs are presented in combination, the animal can suddenly die.

The numbers of interactions increase geometrically with the number of variables involved. When only temperature and humidity are considered, the only possible interaction is temperature by humidity. However, normal environmental situations may involve a multitude of components of the weather matrix changing at various times. When changes in temperature, humidity, barometric pressure, wind speed, and ionic concentration occur, the number of potential interactions is $5 \times 4 \times 3 \times 2 \times 1$ or 120. The smaller interactions, such as a barometric pressure by ion concentration interaction, are relatively easy to duplicate in the laboratory and to test under controlled conditions. The larger interactions, such as those that occur only when all five factors are changing, become almost intractable. However, theoretically at least, they may still evoke specific behavioral-biological changes.

Other factors may contribute to or alter the weather matrix. When they are immediate consequences of weather changes, such as release of soil gases or the movement of pollen, they do not complicate the analysis. However, factors with slower periods of occurrence may be superimposed upon the weather matrix. For example, monthly or yearly changes in solar activity or the physical consequences of lunar phase can add to the complex interactions so that some bizarre or odd disturbance in human behavior occurs. Without that last factor, in principle, the interaction would not be complete and the behavior would not occur.

CANDIDATE STIMULI OF THE WEATHER MATRIX

The following environmental factors have been implied by correlational analyses or have been demonstrated experimentally to produce behavioral and biological changes similar to those associated with weather in general. Most will be discussed in separate chapters.

Temperature

Ambient temperature changes that influence the transfer of heat energy between an organism and its external environment have been popular explanations for most biometeorological effects. Temperature changes have great simplistic and theoretical appeal since the fundamental requirement for mammals to maintain internal (core) temperatures at 37° Centigrade (°C) or about 98° Fahrenheit (°F) is well known.

Relatively sudden and lasting changes in ambient temperature alter the rate at which heat is lost or gained across the body surface. The consequent change in heat concentration in the body boundaries, such as the skin, stimulates receptors that in turn send information directly to appropriate control centers in the brain, especially the hypothalamus. This portion of the brain not only responds reflexively to sudden changes in heat or cold but also stimulates other portions of the body.

Stable decreases in external temperature evoke gradual constriction of the blood vessels in the skin. This reduces the total surface area of blood exposed to the decreasing temperature and helps reduce heat loss. Construction of blood vessels can also be accompanied by increased adrenalin output from the adrenal glands above the kidneys and by alterations in the body electrolytes. The net consequence of the peripheral response in the skin not only influences heat loss, but also modulates general motor activity, such as ambulation, and urinary output.

The hypothalamus is the key brain structure to integrate emotional expression (but not experience). A vast array of autonomic or vegetative behaviors, such as aggression, sexual activity, respiration, heart rate, eating, water consumption, fluid balance, and internal temperature balance are controlled by this structure. Excessive changes in temperature of the body boundary could indirectly affect these behaviors as well. The human being would not necessarily be aware of these weather-related sources of behavior. More likely, the behaviors would be explained in the context of a more immediate and obvious stimulus such as a social event.

If functionally adjacent portions of the hypothalamus, such as the amygdala or septum, were stimulated, the weather effect would not be a "tip of the tongue" experience. Instead, the diffuse, non-specifiable experiences would acquire a twilight of "meaningfulness." The person may actually report feelings of irritability, mood alteration, or difficulties in concentration.

Indeed, parts of the brain that respond to temperature are also involved with most of the behavioral patterns known to be correlated with weather. Experimental stimulation of the human hypothalamus, amygdala, and septum with electrodes can induce euphoria, depression with suicidal impulses, aggression, sexual activity, agitation, lethargy, and difficulties in concentration. However, these portions of the brain also appear to respond to a variety of psychological and social situations as well as to other stimuli of the weather matrix. Many of these stimuli are coupled intimately with temperature changes.

Temperature change is a likely source of weather-related behaviors when human systems are directly exposed to its consequences. However, there are reports of similar behaviors from individuals who presumably live in temperature controlled conditions. Some other component of the weather matrix may have evoked their behaviors.

Humidity

Humidity is a measure of the moisture or water vapor in the air. The most common measure is relative humidity, measured in percent units. Relative humidity reflects the percent of the total capacity of water vapor a given parcel of air can maintain before condensation begins to occur. For example, a relative humidity of 50 percent indicates that the air contains only half of the total water vapor that it can maintain before saturation and condensation.

Since the mixture of water vapor with other atmospheric constituents follows the same chemical principles as other gas interactions, the amount of water vapor that can be contained within a parcel of air is dependent upon temperature. Warmer air can contain more water vapor than colder air, similar to the potential for a warm glass of water to dissolve more salt than a cold one.

parcel of air at 5°C with 90 percent humidity would contain less total water vapor than one at 35°C with 90 percent relative humidity.

Numerous everyday examples demonstrate the effects of temperature upon the amount of water vapor within air. When the surrounding air is around 30°C and the relative humidity is high, such as 80 percent or more, colder objects appear to collect water droplets. Essentially, the air around these objects is cooled beneath its potential to hold the amount of water vapor, and water condenses upon the object. Similar processes occur when warm, humid air is cooled locally due to the absence of a heat source, such as during the night. Below a critical temperature, the water vapor can no longer be maintained within the air mass, and it condenses as dew or frost upon surrounding objects.

The relative water vapor in the air mass has great biometeorological potential since the temperature and concentration of water vapor influence the heat loss from the organism's body surface. Around biologically optimal heat exchange temperatures, between 60°F and 80°F, low humidity (10 percent) facilitates the evaporation of water from the external (skin) and internal (lungs) body boundaries exposed to it. Consequently, the nose, throat, lungs, and skin become drier. Since the body is already around 80 percent water by concentration, moderate humidities at these temperatures have less severe consequences.

Above 90°F (32°C), high humidity (above 80 percent) progressively interferes with the body's heat exchange with the surrounding air. Blood vessels in the skin must dilate more to release heat produced by the normal consequences of biochemical activity. Even mundane activities, such as walking, climbing stairs, or prolonged standing, that require comparatively small amounts of biochemical activity can result in copious perspiration. During excessive vasodilation, blood pressure drops and the person may experience dizziness or black spots in the visual field.

Sudden influxes of cool and very humid air significantly accelerate the rate at which heat is lost from the skin and lungs. The increased water content of the cold air acts like a heat sink. Heat energy from the body is absorbed by the cold water vapor, resulting in relatively quicker drops in peripheral body temperature. Even when the hypothalamic centers compensate for these losses, the humid, cool air continues to quickly absorb biologically generated heat energy.

Barometric Pressure

Barometric pressure is an indicator of the force per unit area exerted upon the living organism, usually adjusted to sea level, by the overlying atmosphere. Barometric pressure has been measured by many units including millimeters (mm) of mercury, millibars (mb), and kiloPascals (kPa). On maps, areas of equal barometric pressure are joined together by lines called isobars. When drawn appropriately, isobars give a pictorial concept of air masses.

Barometric pressure is a function of the number of air molecules within a unit volume of air. Since the number of molecules within a given volume of space is dependent upon the amount of heat energy (thermal agitation), it should not be surprising that barometric pressure and temperature are interrelated. Air molecules within cold air masses have less thermal energy and require less space in which to move. Consequently, more molecules occur within a given volume of space, and the net barometric pressure is higher.

Within a warm air mass, molecules have relatively greater thermal energy. The greater vigorous movements of the air molecules increase the distance between the molecules, resulting in fewer molecules per unit volume. Consequently, warmer air is associated with lower barometric pressures. This assumes no adjacent boundary exists.

Changes in barometric pressure have been correlated with several biometeorological phenomena, ranging from heart attacks to arthritic complaints. The popularity of barometric pressure as the explanatory variable of weather-related behaviors is especially evident among amateurs. However, the physiological mechanisms by which barometric pressure influences human behavior are very unclear. Often, unfortunately, this situation facilitates the generation of nonsense disguised as scientific theory.

Wind

Wind velocity, as measured in miles per hour or kilometers per hour (km/h), can be an indirect measure of the relative speed of energy exchange between adjacent air masses. Wind direction, on the other hand, can be used to isolate the relative compass direction of the low pressure region generating the wind. For example, with the wind at your back, the center of the surrounding low pressure system is to the left (in the northern hemisphere).

Wind speed has organismic relevance since it enhances the effects of temperature and humidity. Most people in the northern temperate zone are familiar with the chill index, which combines temperature, humidity, and wind speed into the chilling capacity of a parcel of air. Temperature and humidity conditions that would otherwise evoke a slightly unpleasant effect can induce severe frostbite when the wind factor is added. Wind velocities also influence the diffusion of outside air through the walls of dwellings, thus increasing the rate at which the biological system within is exposed to weather changes.

The presence of wind in the weather matrix contributes to other potential stimuli. Wind facilitates the movement of electrically charged ions, microscopic particles, pollen, and bacteria. Whereas normal diffusion of these materials in the air would be relatively slow, wind increases the concentration of particles to which an organism is exposed. At moderate to severe wind velocities, the effects of both audible and nonaudible (infrasound and ultrasound) sound components, become potential biological stimuli as well. Special winds, such as the Foehn in Bavaria or the Minstrel in France, are traditional explanations for human aggression and illness.

Ions

Atmospheric ions are portions of molecules or conglomerations of particles that have acquired a net positive or negative electric charge. According to convention, positive ions have insufficient numbers of electrons, while negative ions have too many electrons. This absence of charge neutrality gives ions significant reaction potentials. There is a constant tendency to reduce the charge disparity—to become neutralized—by borrowing other electrons, in the case of a positive ion, or by losing excess electrons, in the case of a negative ion. The source for these exchanges can exist within biological tissue as well as within atmospheric constituents.

Since ions are charged, they are influenced strongly by the direction and intensity of surrounding electric fields. The direction of an ion's movement within three-dimensional space is a function of its charge and the direction of the local field. The velocity of an ion's movement through space is a function of its size, charge, and the intensity of the electric field. Small ions, such as ozone (O_3^-), move at about one centimeter per second (cm/sec) in an electric field of 1.0 volt per centimeter (V/cm), while larger ions, called Aitken nuclei or Langevin ions, move only about 0.0001 cm/sec in the same intensity field.

Because of their excess charge, a basic unit of chemical reactions, atmospheric ions have the potential to influence normal biological activity. The intensity of the effect would no doubt reflect their concentration as well as their access to key biochemical pathways. Although atmospheric ions could certainly affect the skin surface, a theoretically more potent mode of interaction would be through the nostrils and lungs. Appropriate ion concentrations taken into the lungs could easily be dissolved into body fluids. The lungs, which are essentially an internal body surface exposed to external stimuli, contain a vast surface area of absorbent tissues, numerous blood vessels, and reactive cells.

The concentration of ions has been correlated with many of the behaviors associated with barometric pressure. Positive ions have been implicated as the culprits responsible for undesirable human conditions, such as aggression, suicides, homicides, and infections. Negative ions have been considered in a more benevolent light as a source of improved concentration and feelings of well-being. In fact, there have been occasional fads to breathe air passed through negative ion generators in order to improve behavior. As shall be seen, this simplistic dichotomy of good and bad as a function of charge can be erroneous.

Electric Fields

Electric potentials, usually measured in volts per meter (V/m) or volts per centimeter (V/cm), are generated when a disparity of charge exists across space. If more electric charge exists on one side of a boundary, a potential difference exists across the boundary. If the boundary were removed, the charge distribution would become homogeneous, for all practical purposes across that space, and the potential difference or voltage would drop to zero.

A similar situation exists on a global level. Significant electric potentials occur between the earth's surface and the upper atmosphere, which contains relatively high concentrations of ionized particles. Since air is a comparatively poor conductor, the disparity of charge and hence the voltage between the earth's surface and the upper atmosphere is maintained. If air were a good conductor, like copper, the potential difference would not exist since the charged particles would have flow access to the ground. However, the air is not a perfect dielectric (a medium that separates charge) either, and there is some leakage of charge toward the ground, called the earth current.

During clear weather conditions, the typical potential gradient between the earth's surface and some vertical distance is about 130 ± 20 V/m (that is, daily range is from 110 to 150 V/m). This means that between the surface and a distance of one meter high, a potential difference of 120 volts exists; between the surface and a distance two meters high, a difference of 240 volts exists, and so forth. Since this difference is not concentrated into a small area, like household outlets, and because of the low current flow (small amperage), normal intensities contain very limited biopotent potential.

The potential gradient can change. Any process that alters the relative ion concentration in, or the distance of the upper atmosphere above, the surface will influence the potential gradient. Consequently, clear weather electric gradients change as a function of time of day, season (distance from the sun), and solar disturbances. A much greater change in the local potential gradient occurs during brief, localized weather disturbances. Thunderstorms can generate potentials of more than 10,000 V/m. If the air is sufficiently dry, the voltage may discharge from the tips of leaves or other sharp objects (coronal discharges). Other weather conditions, such as falling snow or fog, can generate potential gradients greater than 1,000 V/m.

Changes in the electric field strength beneath various air masses are known to influence insect behavior and even to alter the clotting capacity of some fluids, including blood. The mechanisms by which these reactions take place and their relevance to human behavior are not clear. Theoretical possibilities will be discussed later.

Electromagnetic Fields

In nature, electromagnetic (EM) waves can display an unbelievably vast range of known frequencies. X rays, ultraviolet waves, light, infrared (heat), microwaves, radio waves, and Schumann waves are well known examples of specific EM frequencies or wavelengths. Thunderstorms are frequently associated with EM waves in the very low frequency (VLF) range, that is waves between one kiloHertz (1 kHz or 1,000 cycles/second) to several hundred kHz. On the radio, these waves are registered as static.

Moving, unstable weather systems, such as low pressure air masses and especially frontal regions, are associated with VLF or higher frequency waves emitted as ELF (extremely low frequency) pulses. The ELF range usually involves EM waves, fields, or pulses between about 1 and 100 hertz (Hz). ELF pulses and waves around 4 Hz appear to predominate in some localities during cloudy conditions, while higher frequency fields predominate during clear weather.

ELF waves and fields have received only recent attention as biologically relevant stimuli. ELF fields generated in the environment are similar in amplitude (intensity), shape, and characteristics to those generated by the human brain. Theoretically at least, this similarity allows for possible interaction between humans and weather systems. In addition, ELF fields have a 99 percent capacity to penetrate most surface dwellings. Since ELF fields and pulses can be generated hundreds of kilometers away from the unstable weather sources, these stimuli have the potential to evoke organismic changes long before the typical alterations in temperature, humidity, and pressure occur.

INDIRECT CONSEQUENCES OF THE WEATHER MATRIX

As a frontal area or air mass passes over terrestrial space, a variety of indirect environmental agents can be released or emitted. Although always present in the environment, changes in weather facilitate their rates of accumulation or production. These stimuli can also contribute to the total impact of the weather matrix upon biological systems.

Soil Gases

The soil, or approximately the upper one to two meters of surface, behaves like a gigantic vat of chemical reactions. Decaying organic matter, minerals, and various fluids, primarily water, react in a myriad of combinations to release various products. Some of these products are gases, such as hydrogen sulfide (H_2S), carbon dioxide, methane (CH_4), and phosphine (PH_3), to name a few. Under static conditions, escape rates into the air above the ground are constant.

Sudden decreases in barometric pressure, especially when accompanied by very warm air, increase the rate at which the gases are released into the air. At sufficient concentrations, these gases, which are always present in some small concentrations, become apparent to the living system. The unpleasant olfactory consequences of hydrogen sulfide, often called sewer gas, have direct effects. At higher concentrations, H_2S can be toxic. Methane gas is frequently released around swampy areas; when combined with phosphine, the will-o'-the-wisp is formed.

Organic Materials

Sudden increases in wind speed associated with air mass approaches can release greater amounts of pollen spores or micro-organisms from source areas. The potential pool of total organics is dependent upon season. The same weather matrix in one season may not be associated with certain behaviors during another. This problem makes simple correlational analyses inappropriate.

Sudden increases in pollens or spores in the air trigger allergic reactions including skin irritation, asthma, sneezing, and sinus congestion. Since allergic reactions often involve systemic (whole body) changes in immunological reactions, the person may report feelings of tiredness and flulike or coldlike symptoms. These, in turn, can be exacerbated by other aspects of the weather matrix.

Man-made Pollutants

The effects of the weather matrix upon the relative concentration of man-made pollutants are evident in heavily industrialized countries. The combustion of fossil fuels emits sulphur dioxide, while light (especially ultraviolet) acting upon petroleum products creates ozone and several metal-organic pollutants. Although ineffective at normal concentrations, their toxicity can be enhanced by the weather matrix.

Smog conditions are frequent occurrences in many industrialized cities, such as London, Los Angeles, and New York. Inversions, whereby the position of warm air is maintained by overlying cold air, are highly correlated with severe smog. Cities that have been built in relative depressions, such as valleys or basin regions, are quite prone to inversions.

Geophysical Effects

The indirect consequences of moving air masses upon potential biologically relevant geophysical stimuli are still theoretical. Low pressure air masses exert less weight upon large areas of terrestrial surface than do high pressure air masses. Since the upper crust of the earth is partitioned into blocks or local plates, variable pressures from overlying air masses can shift their relative positions very slightly. The consequent adjustments along the boundaries of the plate, which may be hundreds or thousands of kilometers in width and length, are associated with microseisms—small earthquakes. They display major frequencies between 0.1 Hz and 10.0 Hz, but have very, very low amplitudes.

Certain species of sea life appear to respond to the micro-vibrations preceding the passage of low pressure air masses. Whether human beings can detect or respond to these weak mechanical vibrations has not been demonstrated clearly. Human beings are composed of matter, and consequently have intrinsic mechanical vibrations. Usually these vibrations are invisible to the naked eye and occur between about 1 Hz and 20 Hz. The potential resonance interaction has been used as the theoretical basis for microvibration detection.

ASTROGEOPHYSICAL CORRELATES OF THE WEATHER MATRIX

To be thorough, each major environmental factor consistently correlated with specific weather conditions must be considered to have at least a partial role in the total organismic response to the weather matrix. Perhaps these roles are so small that they are irrelevant. However, that conclusion must carefully be evaluated. A number of environmental factors have been consistently correlated with unstable weather conditions.

Geomagnetic Storms

In the last decade, global data collection has allowed correlational analyses never before possible. From these data, a new pattern has emerged indicating unsuspected interrelationships within world weather. Global geomagnetic storms, influenced primarily by the sun's activity but to some extent by lunar orbit, appear to be persistently associated with increases in unstable weather masses three to five days later. Since human beings may respond to geomagnetic disturbances, as some crude correlational studies indicate, this factor may elicit responses—delayed in time—that would be erroneously attributed to other weather-related stimuli.

Lunar Phase

The existence of a small but persistent relationship between lunar phase and aggressive behaviors, birth incidence, and psychiatric admission rates is not disputed by knowledgeable scientists. The major argument concerns the mechanism by which these correlations occur. Are they due to purely psychological factors of expectancy, to specific weather coupled to lunar phase, or to unspecified, direct lunar-geophysical stimuli? Most cultures contain a rich lore of alleged lunar-related behaviors—from madness to uncontrollable sexual urges. People raised within these cultures would have been conditioned to respond with the behaviors of the myths. In addition, periods of full moon have been noted by geophysicists to coincide with increased vigor of thunderstorm activity, altered geomagnetic disturbance, and influxes of ions from the upper atmosphere. Each of these factors by itself could contribute to the total biometeorological effect.

Infrasound

Infrasonic stimuli originate primarily from terrestrial sources, although they do occur at very, very low intensities in the upper atmosphere following severe solar storms. Typically, infrasonic vibrations between 0.1 Hz and 10 Hz are associated with volcanic eruptions, distant tornados, and probably all major land movements. They are mentioned in the context of the weather matrix since they may be superimposed upon the local weather matrix.

Solar Disturbances

Since the heat energy for all weather ultimately originates with the sun, it is not surprising that relatively small alterations in the sun's output have been correlated with a vast array of upper atmospheric and weather activities. Geomagnetic storms are primarily associated with solar disturbances that are characterized by the emissions of fantastic energies into space. These emissions are reported as flares, X-ray bursts, or spurts of corpuscular radiation (solar plasmas). Although the latter are always bathing the earth in a continuous medium called the solar wind, periods of increased solar activity produce massive eddy currents and distortions in the solar wind. These perturbations in turn strike the earth's outer atmosphere and magnetic field.

Alterations in solar storm probabilities follow specific periodicities. Predominant periods are associated with the approximate 28-day rotation of the sun and with the sunspot cycle (approximately from 11 to 12 or from 22 to 24 years). For reasons discussed later, the number and the severity of solar disturbances are highly correlated with the number of sunspots produced on the solar surface. Both massive climatic changes on the earth's surface and alteration of human social behaviors have been associated with solar activity.

2

CONCEPTUAL SITES OF ACTION
WITHIN THE HUMAN LOCUS

The human being can be considered a mobile, finite volume
of substances, suspended or dissolved primarily in water, structurally
supported by calcium salts (the bones), and encapsulated by a semi-
permeable boundary on the outside (the skin). This boundary continues into the mouth and nose to include the linings of the lungs and
of the gastrointestinal tract. Within this mobile volume, a vast
variety of reactions occur in response to environmental stimuli.
These reactions contribute to what scientists have labeled physical,
chemical, biological, and psychosocial behaviors.

Given this description of the human organism, one must begin
to systematically isolate areas within three-dimensional biological
space that could respond to weather-generated stimuli. Do stimuli
from the weather matrix affect the whole volume in the same manner?
Do different stimuli from the matrix influence the whole volume in
different ways? Are there particular parts of the three-dimensional
volume that respond more to weather stimuli than do others?

If the distribution of body mass were homogeneous, the analysis
of weather effects would be relatively simple. However, this is not
so. Some materials and reactions cluster within very localized
regions, while other materials and reactions are spread throughout
the entire volume. As a general label, the highly localized reactions
and materials can be called high gradient systems, while the more
diffuse reactions can be called low gradient systems. They are
considered systems since each contains elements that share probabilities of occurrence.

High gradient systems are more well known to the scientist
since they can be easily seen and described. Examples include the
nervous and cardiovascular systems. Low gradient systems are

less easily perceived by the human sensory apparatus and hence have often escaped crude observation. Examples are the total water of the body volume, the ground substance, whole body electromagnetic fields, and organismic mechanical vibration.

Any or all of these systems could respond to the complex stimuli of the weather matrix. If the human organism is considered a physical system as described, one must expect that biophysical principles will be followed in general, although specific responses may differ greatly due to the complicated geometries of application. Whereas changes in temperature would be more likely to influence surface blood flow mechanisms, changes in barometric pressure would be expected, statistically, to influence whole-body fluid dynamics. Since biometeorologists are only in the initial stages of data collection to answer these questions, each potential system must be stated, described, and treated with equal probability.

HIGH GRADIENT SITES OF WEATHER STIMULATION

High gradient sites of possible action by meteorogenic stimuli include the central nervous system, the peripheral nervous system, the vascular system, and particular organs. Upon gross inspection of the human body, these systems are easily seen. They have been popular candidates since they are discrete and easy to measure. Whether meteorological stimuli influence these systems directly due to some special property of detection or indirectly through the many chemical reactions outside their immediate boundaries is not clear.

The Central Nervous System

The central nervous system (CNS) is composed of the brain and spinal cord. As the term implies, the CNS appears to integrate the incoming sensory information from both specialized and diffuse receptors along the body boundaries, to compare it with previous experiences that are represented as various configurations of protein and electromagnetic patterns (memory), and then to respond both outside and inside the body. An external response primarily involves the skeletal muscle: walking, talking, or running. An internal response would involve changes in heart rate, respiration, chemical secretions, and even private experiences or thoughts.

The basic unit of the nervous system is the neuron. Neurons are specialized cells for conduction, as opposed to irritability (sensory cells) or contractility (muscle cells). They have two major modifications or processes of the cell membrane. One type is the dendrite, whose primary function is to receive information from other neurons. Dendrites protrude from the cell body like branches from a tree. The other major process is the axon. There is only one axon for each neuron, but this extension can be thousands of times longer than its width, thus producing a cable property within the nervous system. Although calculations vary, estimates indicate that the human brain contains approximately one billion (10^9) neurons.

Since each neuron can contact hundreds of other neurons, the brain can be seen as a vast net or field of axons and dendrites, interacting, overlapping, converging, and diverging. Over these pathways, transient changes in electrochemical potential occur. These local changes, labeled action potentials or spike potentials, exist for about one-thousandth of a second (10^{-3} sec) and are mediated along the axons at about 100 meters per second (m/sec).

Most neurophysiologists have assumed that the patterns of these discrete changes in electrochemical potential in space/time are the language of the brain. From one perspective, all words, thoughts, sensory experiences, complex perceptual forms, and even concepts of self can be correlated to particular patterns of these changes. Since action potentials occur in an all-or-none manner, brain language can be compared to the digital (all = 1 and none = 0) logic of electromechanical computers.

Direct effects of the weather matrix upon the CNS appear unlikely for both theoretical and practical reasons. First, the brain is well within the body core where small changes in temperature, humidity, or wind velocity cannot directly reach. About the only weather stimulus that might penetrate the skull and vertebrae is ELF electromagnetic or sonic field/waves.

Temporal components are not optimal. The thinking activity of the brain works at time intervals of about a millisecond (10^{-3} sec), while typical weather changes occur from hours to days. Again the only direct weather factor that falls within the range of brain language is ELF stimuli. The impact of its effects has not been well demonstrated.

Without further theoretical deliberation, a more direct proof is the general lack of large organismic effects produced by weather changes. Since one does not see large variations in the content of talking, massive changes in the content of thinking (as reported), or disruptive alterations in the actual sequences of human behavior, it is unlikely that weather directly influences the brain. Otherwise weather would have been associated long ago with hallucinations, varied experiences to the same stimulus, and unbelievably massive disruptions in human social behavior. Since human experience is closely correlated with the sequence of action potentials and the actual detailed patterns display great individual variation, direct disruption would have catastrophic consequences. The world would be a hallucinatory jumble of experiences.

Even low-level, direct interference with the brain's language would show statistical effects. Although small changes would produce only occasional or very small alterations in experience—perhaps not even detectable in the individual—their occurrence would become obvious when a population was evaluated. At this level, weather changes would be associated with statistical increases in misinterpretations of ordinary stimuli, such as perceiving a car as a truck, or frank hallucinations.

There is no clear evidence to indicate such profound effects of weather upon human beings. However, the nervous system does contribute an important component to the total weather effect: memory. The CNS, although not directly affected by weather per se, can be conditioned by repeated pairings of initially ineffective sensory input with stimuli that induce changes in biobehavior. This capacity for conditioning or learning and memory is often a major source of misleading evidence of biometeorological mechanisms.

The most essential component of conditioning is space/time contiguity: the pairing of stimuli. For weather-related effects, the experimental procedure would involve some form of classical or respondent conditioning. In this situation, some unconditioned stimulus (UCS) elicits an unconditioned response (UCR). The capability for the UCS to evoke this change is usually genetically based: it has not been learned.

When a neutral stimulus that does not elicit the UCR is presented in space-time with the UCS, conditioning can occur. By merely presenting the neutral stimulus and the UCS together several times, the neutral stimulus soon develops the capacity to elicit a response very similar to the UCR. When this happens, the neutral stimulus is called a conditioned stimulus (CS), while the response elicited is called a conditioned response (CR).

Classical or respondent conditioning does not require active participation by the organism; in fact, the organism can be quite passive or restrained. For example, if seeing the clock "with the big hand on the 12 and the little hand on the 5" is frequently followed by eating dinner, just looking at the clock can initiate hunger pangs or anticipation of food. Yet, by itself, without the conditioning, 5 o'clock may not evoke any particular effect. Many forms of incidental learning, from anticipating a sequence of songs on a tape deck to misinterpreting a prediction about a friend's behavior as so-called ESP or telepathy, are frequently a consequence of this form of conditioning. Contrary to popular beliefs, the human being does not have to be aware of the situation in order to be conditioned.

In a biometeorological context, passive learning or classical conditioning is a very important source of causal misinterpretation. For example, there may be nothing depressing about the color gray or low barometric pressures. But when paired accidently or due to some constant correlation with another stimulus that induces depressive and bloated feelings, these two stimuli are able to evoke depressive and bloated feelings as well. A scientist not familiar with the learning history of a person or with the phenomena of conditioning in general might erroneously conclude that gray, low pressure days cause depression.

The Hypothalamus: Emotional Integrator

The hypothalamus, although part of the CNS, deserves special attention in biometeorology. There is nothing anatomically conspicuous about this lower middle portion of the brain. Except for its close proximity to the third ventricle and to the pituitary, one would not expect such far-reaching functional implications.

However, although we see nothing remarkable about this structure with the naked eye, it is an important site of meteorotropic interaction. The hypothalamus is packed with axons from different chemical systems throughout the brain. It is primarily involved with the same types of behaviors frequently correlated with weather changes. For example, the hypothalamus appears to integrate the behaviors of eating, drinking, sexual activity, aggression, body temperature, and even general activity. It may also alter disease susceptibility.

Hypothalamic inputs appear to originate all over the body and brain. A primary source of input comes from the autonomic nervous system, described in the next section, which is in close contact with body cells. The hypothalamus is in a unique position to influence both internal and external responses. Its connections to one part of the pituitary, the neurohypophysis, controls the release of the anti-diuretic hormone vasopressin, involved with fluid retention and water balance, and oxytocin, involved with contraction of the uterus, especially during birth, and milk ejection during lactation. Connections with the other portion of the pituitary through releasing factors allow the hypothalamus to influence the thyroid, adrenal glands, sex glands or gonads, and general cell metabolism.

External responses have been demonstrated in nonhuman animals, most frequently by electrically stimulating some part of the hypothalamus. Full-blown rage, fear, fighting, and sexual behavior have been displayed. In human subjects, stimulations of the hypothalamic regions produce similar gross behavioral alterations. If the stimulation, due for example to a tumor or chemical disturbance, is too excessive, functionally connected portions of the hypothalamus— such as the amygdala, hippocampus, septum, anterior thalamus, cingulate gyrus, and frontal orbital cortexes—are excited, and the person may report experiences of mood or emotional alteration. The details of the experience will be couched within the individual episodes of each person's experience.

There is an abundance of experimental evidence to support a major hypothalamic role in weather responses. Its general architectural position allows access to massive incoming sensory input since it is below the thalamus, the major sensory-motor integrating region before awareness occurs in the cerebral cortex. Its microarchitecture shows regions where small blood vessels, such as capillaries, are highly concentrated. The paraventricular nuclei and supraoptic nuclei of the hypothalamus control the release of oxytocin and vasopressin, respectively, and are among the most densely capillarized portions of the human brain. Such high density capillarization allows enhanced interaction between functional neuronal surfaces and extremely small alterations in blood constituents.

The sensitivity of the hypothalamus to minute physical and chemical changes in the blood is extraordinary. Early experiments indicated that changes in central blood temperature of only $0.01°C$ evoke detectable (0.25 milliliters per second) alterations in cutaneous (skin) blood flow. Cooling of the anterior hypothalamus through direct experimental means is associated with peripheral vasoconstriction and changes in blood flow within one to five seconds; warming the same region evokes opposite effects within a similar period. These vascular responses can be stimulated by direct neuronal connections or indirectly through the release of the vasomotor compounds: adrenalin and noradrenalin from the adrenal glands.

Since the hypothalamic nuclei are like clusters of set points that compare various physical/chemical inputs to some present level determined by experience or genetics, direct damage to them can produce exaggerated behaviors. Damage to the anterior nuclei, where the set point for $37°C$ body temperature is maintained, may result in the human being actually freezing to death in a normal room temperature. Damage to the nuclei concerned with glucose (blood sugar) levels may result in constant nibbling so that the person never seems to get enough food and becomes obese or is persistently hungry. Examples could be given of many other behaviors controlled by the set points of the many hypothalamic nuclei.

Direct stimulation of critical hypothalamic areas produces clear and blatant behaviors that override the thin veneer of psychosocial constraints. Consequently, even considering ELF sonic and electromagnetic stimuli, it is unlikely that the weather matrix directly stimulates the hypothalamus. This brain region does, however, have a central role in the integration of autonomic activity, changes in blood characteristics, skeletal muscle changes related to emotions, and relationships with the body boundaries. The hypothalamus forms a nexus among parts of the brain correlated with awareness. As with all integrators of complex function, very, very small changes in its operation could produce transient but potent changes in behavior.

The Autonomic Nervous System

The autonomic nervous system comprises the neurons and their axons and dendrites outside the brain and spinal cord that innervate the internal organs, the blood vessels, and the body boundaries. Functionally, this system has been associated with the initial emotional reactions or "gut reactions". This system is more difficult than the central nervous system to study anatomically since most of the neurons are clustered in various-sized areas (plexi) that are dispersed throughout the viscera. These clusters of neurons are very complex perceptually, often making "no sense," and may vary in number between individuals.

Autonomic neurons induce and detect changes within their target organs of the viscera. Organ-specific alterations in blood flow, activity, and probably select chemical reactions are coded into neuronal language and then relayed primarily to the hypothalamus through portions of the brain stem. At this level, reflexive alterations occur to adjust the particular organ to some set point, much like a thermostat responds to changes in room temperature by increasing or decreasing heat output in order to maintain a set temperature. When the stimulation becomes too intense, the person may become aware that something is wrong. However before becoming aware, an entire continuum of changes can influence human behavior.

The autonomic nervous system is composed of two integrated but antagonistic nerve fields, designated as orthosympathetic (or just sympathetic) and parasympathetic. Each nerve field has characteristic general patterns of distribution to the target organs and a characteristic chemical transmitter between its axons and the target organ. The transmitter in the sympathetic nervous system is noradrenalin, the immediate precursor to adrenalin, while in the parasympathetic nervous system the transmitter is acetylcholine. Effects mediated by each of these systems are called adrenergic or cholinergic, respectively.

In general, sympathetic activity is associated with arousal and activity of the whole organism (the classic alarm state), while parasympathetic activity is associated with relaxation and passive conditions. Each system's affect upon a target organ is mutually antagonistic. For example, stimulation of sympathetic fibers distributed to the heart produces increased heart rate, while stimulation of parasympathetic fibers to the heart decreases heart rate. However, inhibition of normal parasympathetic activity to the heart can also increase heart rate.

Most central visceral organs receive fibers from both para- and orthosympathetic systems. The particular consequence of stimulation from either of these systems is related to the particular function of the organ itself. Thus stimulation of sympathetic fibers innervating an organ that evokes activity produces activation, while sympathetic stimulation of an organ that normally inhibits initiates inhibition. Some of the organs innervated by the autonomic nervous system are dilator/constrictor muscles of the eye, lacrimal (crying) glands, the salivary glands (parotid, sublingual, and submaxillary), the heart, lungs, liver, pancreas, spleen, stomach, small intestine, colon, kidney, bladder, genitals, and a multitude of blood vessels. All are stimulated directly or indirectly when general sympathetic or parasympathetic activity occurs.

When general sympathetic arousal occurs, the following changes are typically evident: dilation of pupils; construction of capillaries and arterioles in skin and abdomen (depending on the receptor), but dilation of these vessels in the heart and skeletal muscles; relaxation (dilation) of bronchi in the lungs; increased heart rate, blood pressure, and stroke volume; increased oxygen consumption, but decreased oxygen use; elevations of blood glucose, calcium, and acidity levels; faster clotting time of blood; and increased frequencies of cerebral cortical electrical activity.

Orthosympathetic arousal induces both general and situation-dependent behaviors. The person becomes more active and talkative and may feel strong, full of pep, and sometimes even euphoric. Subjectively, the person may display more egocentric behaviors, such as enhanced self-appreciation or self-confidence. Individual learning can contribute to the type of psychological and behavioral responses. If general sympathetic arousal has been associated with aggression, the person may feel more aggressive during these periods.

Interestingly, elevated concentrations of adrenalin in the blood—a natural consequence of sympathetic stimulation—do not appear to determine specific emotions. Rather, emotional tones are influenced by the cognitive or psychological aspects of the situation. Human subjects injected with adrenalin and placed in situations in which people are laughing report experiences of frivolity and humor. Human subjects injected with the same amounts of adrenalin and placed with people crying demonstrate enhanced experiences of sadness, depression, and related tones. Apparently, an otherwise ineffective psychosocial environment can profoundly influence emotional experience once adrenalin is injected or the sympathetic nervous system is mildly activated.

When general parasympathetic activity dominates, the following occur: contraction of pupils; decrease in blood pressure and heart rate; decrease in oxygen consumption, but increase in its use; constriction of muscles in the bronchi; increased secretions from various glands, such as insulin release from the pancreas and gastric secretion from the stomach lining; decreases in blood glucose, but increases in blood potassium content and blood alkalinity; decreased electroencephalographic (EEG) frequencies; and feelings of tiredness, sluggishness, or general malaise.

Most people show an autonomic balance such that neither system predominately influences a given organ continually. During the day, the average person is sympathetic dominant and consequently displays low level general sympathetic activation. Consequently, heart rate is faster, general activity, both muscular and mental, is greater, body temperature is higher, and the bronchi are more dilated.

During the night, however, the average person becomes parasympathetic dominant. Body temperature is lowered, bronchi are more constricted, gastric and lacrimal secretions are enhanced, and blood pressure is lower. Complications that would be minimal during daily sympathetic dominance become more evident if present. Bulging sensations and pain from ulcers of the stomach lining become more intensified, breathing can become more asthmatic, and cold sweats (from excessive dilation of peripheral blood vessels and lowered body temperature) become more apparent.

The day/night dichotomy between sympathetic and parasympathetic dominance averages over hours. There are several periods in a normal day during which the other system dominates for a while. People get sleepy—parasympathetically dominant—after eating a big meal. After any kind of vigorous sympathetic activity, such as a fight, argument, or sex, a person may feel tired, depressed, or guilty, depending upon a variety of learned factors. These switchover periods during which one system dominates have important ramifications for biometeorology.

Preliminary information indicates that the ortho- and parasympathetic nervous systems do not respond with similar vigor to warm and cold air masses. Skin tests using color changes following chemical application as the primary measure indicate that the autonomic nervous system is more active during the passage of weather fronts. Response latency to cold fronts is less than to warm fronts. Parasympathetic activity becomes noticeable from five to six hours before the passage of a cold front, while sympathetic activity becomes more apparent from three to five hours after the passage. This increased stimulation may continue for as long as from 16 to 30 hours after the passage. While both warm and cold fronts influence the sympathetic system, warm fronts appear statistically to influence the parasympathetic nervous system more vigorously.

The Cardiovascular System

The cardiovascular system includes the heart, arteries, veins, and a myriad of capillaries that innervate the body. Throughout the maze of variously sized tubes of vascular tissue, multipotent blood is delivered to the cells of the body. Some of the tubes can change shape quickly since they are surrounded by smooth muscle, while others remain more or less the same diameter. Visually, the cardiovascular system is seen as a mass of tubes originating from the heart and proceeding along the limbs or into the viscera. From the larger vessels, smaller tributaries emerge from which emerge even smaller vessels.

This pattern of smaller and finer diameter vessels emerging from larger sources continues until one passes beyond the field of normal vision into the microscopic dimension. At this level, tissue is seen to be permeated with a rich nexus of capillaries, arterioles, and venules. Since all cells derive oxygen and nutrients from the blood, tissue appears packed with these components. Some tissues, such as the brain, thyroid, and adrenal gland, have immense volumes of blood vessels compared with their sizes. These tissues are extremely sensitive to minute changes in blood constituents.

The normal tone of the vascular system is maintained by both nervous stimulation from the autonomic nervous system and by the secretion of adrenalin and noradrenalin from the adrenal medulla. Although weather stimuli, especially temperature changes or net temperature alterations enhanced by wind chill factors, could alter heat loss from the blood at the body boundaries, the primary effect would still be mediated through the autonomic nervous system and hypothalamus. Unless weather stimuli have some special effect upon the fabric of blood vessels, any effects would be indirect.

The Blood

Even though the blood is the kinetic component of the cardiovascular system and contains a multitude of different chemicals, it is still contained within a definite locus characterized by high gradients. Blood contains leukocytes (white blood cells), triglycerides (fats), cholesterol, electrolytes (such as sodium, potassium, phosphorous, and calcium), albumin, globulins (related to immune reactions), cell nutrients, and a variety of special proteins called enzymes. Erythrocytes or red blood cells, the most abundant cell type (about 25 trillion in the human body), transport oxygen and gases between the lungs and tissues.

The blood is the major carrier of specific nonneural information to the various organs all over the body. This important substance is the major pathway by which the endocrine glands are functionally connected to higher integrating organs. For example, the pituitary releases very small amounts of hormones—small peptides that are chains of amino acids—that are carried by the blood to particular organs. Specific hormones from the pituitary are known to induce specific effects in the thyroid, adrenal glands, gonads, and probably the skin.

Although the blood volume does not move at the speeds of neuronal activity, it is not a sluggish system. Chemicals injected into the blood from one organ can reach all other organs innervated by rich blood supply within a matter of seconds. Since the whole of the blood in the body passes through the lungs every 30 seconds, substances taken into the lungs, such as pollutants, aerosols, ions, or cold air, could be introduced into the blood stream and to all of its connections very rapidly. The critical features are the substances' solubility and relative concentration.

Direct effects of weather upon the blood have been argued for centuries. Metaphoric descriptions, such as blood becoming thicker before thunderstorms and attributing energy during spring ("spring fever"), sexual activity, or suicidal displays to "something in the blood," are traditional.

Direct effects of weather upon blood, with the exception of changes due to heat loss, would be very difficult to demonstrate. As should be evident, blood chemistry changes are mediated by direct nervous stimulation as well as by organ activity. The number of factors that influence, for example, the formation and destruction of red blood corpuscles are interrelated. Finding the cause is much like trying to find the beginning of a circle.

The Bones

The bones and cartilages are rigid types of connective tissue. Bones are composed of cells and an intracellular field made up of organic and inorganic materials. The primary organic material is collagen and an amorphous protein-sugar material.

The inorganic components of bones are responsible for their rigidity and may constitute up to 60 percent of the dry weight of the skeletal system, depending upon dietary conditions. The inorganic materials are primarily calcium phosphate and calcium carbonate with smaller amounts of magnesium, hydroxides, fluorides, and sulfates. The exact composition can vary with age.

Bones compose the mechanical support for the human locus. They are also a reference for the attachment of muscles for which they act as levers. The central portion of bones contains a soft tissue generally described as marrow. Marrow is the chief blood-forming organ of the adult body. It is considered the single source of red blood cells and granular leukocytes in the adult.

Since bones contain well-developed matrixes of blood and nerves, both direct and indirect stimuli to the bones can alter blood calcium levels and red or white blood cell numbers. Interestingly, the organization of collagen fibers and inorganic crystal structures in bone produces piezoelectric characteristics. Consequently, mechanical distortions or pressures can generate electric phenomena. Whether subtle contractions of large muscle masses, due to cold temperatures or mechanical vibration, could influence bone functions in this manner has been suggested but not demonstrated.

Particular Organs

Final candidates of the high gradient systems of meteorotropic interaction are the organs including the skeletal muscle mass. Sensitivity of one particular organ to weather would necessitate some special physical chemical property of the organ, such as extremes in cell density, cell division, cell heterogeneity, degree of connective tissue innervation, relative concentration of homogeneous materials (such as colloids), and relative concentration of a chemical with a meteorotropic responsiveness.

Since most visceral organs are protected by the body boundary and homeostatic thermal/water balance mechansims, stimuli of the weather matrix must influence organs indirectly. The major exception would be any stimulus that can penetrate the body, totally alter the characteristics of the blood, or induce changes in relative fluid dynamics. There are few such stimuli.

LOW GRADIENT SYSTEMS

Low gradient systems through which weather changes might influence the human organism are more difficult to comprehend and to measure. Since the basic elements of these systems are invisible to the naked eye, they must be measured indirectly through sophisticated instrumentation. The constituents of these systems are complex to most people since physical and chemical concepts rather than common sense activities are involved. Observation of what appears to be a homogeneous mass of "something" throughout the body volume can be disconcerting.

Comprehension of mechanisms by which diffuse changes pervade the body volume becomes imminently complex. For example, how would the statistical alteration of the shape of a molecular species spread throughout the body produce specific changes in behavior? Are these changes effective because the minute changes are integrated over the entire volume by a central nexus, such as the autonomic or vascular system? Are these changes effective because different organs of the body contain disproportionately more of the critical molecular species such that when it is influenced the particular organ is influenced more as well?

The latter possibility implies an interesting conception of the human volume. Essentially, the organs of the body would have special concentration and boundary conditions within which ordinary, elementary constituents of the body volume amass. Incremental changes within these constituents, undetectable at the gross level, would be statistically enhanced in areas of high density. From this perspective, the human system is not a biological organism composed of specific colored organs with different functions, but instead a mobile volume containing probabilistic distributions of special mass.

The Water Molecule

Assuming a simplistic and homogeneous source of weather-related changes, the first logical candidate must be water. The human volume is composed of from about 70 percent to 80 percent water by weight. Some theorists have argued that small changes in the water molecule's physical shape would alter major enzymatic reactions. Since so many chemical reactions are immersed within a water environment, the changes would be mediated in a chainlike reaction across the various levels of organization: chemical, cellular, organ, and behavioral.

Descriptions and, more importantly, empirical support of the mechanism by which such alterations could occur are less rigorous than supposed. Most theorists implicate the infamous 105° bond angle between the two hydrogen atoms attached to the oxygen atom. The means by which this is accomplished usually involves some electromagnetic stimuli. Often the reader leaves such theories with the impression that the mechanism question has been substituted by a semimystical "vague causes vague" argument.

Like the other diffuse gradient candidates, small changes in bond angle could be argued semantically—that is, with words, not with data or equations—to support weather-related mechanisms. If these small changes resulted in a type of general disequilibrium, the major mood would be "quietly uncomfortable," "mild agitation," or perhaps apprehension. If the person had been reinforced for aggression, he would behave this way more often, such as by having more accidents or social assaults. If the person had been reinforced for feeling depressed, increased suicidal thoughts, psychiatric disorders, or general affective disorders would be evident.

The Cell Membrane

Each of the approximately 100 trillion cells of the human body is surrounded by a phenomenon known as the membrane. Although only approximately one one-hundred-millionth of a meter (10^{-8} m) wide, this phenomenon is fundamental for all known life systems. The membrane interferes with the normal powerful kinetics of concentration gradients and charge distributions that would diffuse a given chemical species throughout the body volume. Instead, some chemical species are kept outside the membrane, while others are preferentially stored within the cell body. As a result of this segregation, electrical potentials are generated, chemical concentration energies are displayed, and the phenomenon known as life emerges. If the cell membrane properties were altered, the selective permeability of each cell would be lost and cellular constituents would blend into an oblivious homogeneous soup.

The cell membrane is neither a solid nor a liquid; instead it exists as a liquid crystal, which displays an organized structure (like a solid) and a capability for fluid dynamics (like a liquid). The precise architecture displayed by a cell membrane is a function of the relative water concentration, the local pH (relative balance of hydrogen ions and hydroxyl ions), and temperature.

At life temperatures, the most typical liquid crystal structure of the membrane is a lamelar (sandwichlike) arrangement of phospholipids and proteins that display a selective permeability to chemicals outside the cell. Selective permeability is often conceived as holes within the membrane through which some ions can enter. At the level of the membrane, "holes" become an empty metaphor. Distribution of charge densities is a more likely mechanism.

Changes in pH or local fluid concentration can change the structure of the membrane, allowing chemicals that would otherwise be kept out to enter. At the more extreme limits, the liquid crystal may actually change from a sandwichlike shape to a miscelle (doughnut) shape. When this occurs, all the uneven distribution of chemicals moves toward homogeneity. Essential chemicals move in or out of the cell, and the cell dies.

Between the optimal liquid crystal phase and a phase associated with death are a large array of intermediate conditions. Meteorological stimuli could mediate their effects through these conditions, such as producing statistical increases in the number of leaks within cell membranes. Total loss of only a few cells in the system or just a few leaks within all cells of the system would be compensated by cellular redundancy or by individual cell equilibrium mechanics, respectively. As the meteorological effect becomes more intense, these changes become more pronounced until the initial stages of total system disequilibrium became apparent. These subtle, diffuse effects are monitored by the autonomic nervous system.

The Ground Substance

Most of us have spied a cell in a tissue cross section when looking through a microscope. We have labeled the cell body and its processes as well as the capillaries and support bodies around it. Like most human beings, we have perceived the figure but ignored the ground. Few students either ask about or remember the vast amount of space around each cell, especially those in tissues such as the brain.

Interstitial space is the fluid volume around each cell of the body, even those that compose the outer boundaries. In addition to water, the primary component, this fluid is composed of ground substance. Histologically, ground substance is considered part of the connective tissue system of the body. At life temperatures, ground substance displays a definite ordered structure. Even water molecules appear to be oriented in directions determined by ground substance.

The chemical composition of ground substance is primarily mucopolysaccharides or "sticky sugars," more recently labeled proteinpolysaccharides. There are two main categories: sulfated and nonsulfated, depending upon interaction with body sources of sulfuric acid. The nonsulfated group includes hyaluronic acid and chondroitin. The sulfated group includes chondroitin sulfates A, B, and C, keratin sulfate, and heparin, the well known anticoagulating agent.

Hyaluronic acid is a viscous fluid with a great propensity to bind water. Like other mucopolysaccharides, it is a large poly-anionic net with which water is associated as well as a variety of important electrolytes, such as calcium, sodium, and potassium. This substance has been considered a central agent for alterations in ground substance viscosity and tissue permeability. For example, the enzyme that breaks up hyaluronic acid, hyaluronidase, markedly increases the permeability of tissue. Substances placed on the skin after this substance has been applied easily diffuse into deeper tissues.

Hyaluronic acid is found concentrated in the synovial fluids of the joints, the vitreous humor of the eye, and the umbilical cord. The chondroitin sulfates can be found in great abundance within the aorta, bone, cartilage, skin, and heart valves. Heparin is found primarily, but not exclusively, in mast cells.

Alterations in these substances can have significant effects at organismic levels. For example, alteration of hyaluronic acid concentrations in the fluids of the joints determines the amount of water contained in this limited volume. Excessive water retention would produce swelling (edema) and painful sensations in the joints. Similar effects would occur in people prone to eye diseases. Added pressure within the eyeball from added water retention would increase the pain and interfere with vision.

Since the ground substance system is so diffuse and spread throughout the body, one would expect small stimuli outside the body boundary to be effective. Even changes in barometric pressure would be a likely candidate. In this case, weather diseases would be ground substance or general connective tissue disorders.

Loose Connective Tissue

Ground substance is the fluid component of the connective tissue system. There are also connective tissue cells. Unlike the concept of a specific organ in which similar cells with similar function are packed into the same space, connective tissue cells are spread throughout the body in a loose fieldlike structure. They form a matrix, which, although apparently thin, would be quite substantial if packed into typical organ densities.

Typical connective tissue cells are fibroblasts, macrophages, mast cells, plasma cells, and occasionally blood cells. Fibroblasts are responsible for fiber formation in the connective tissue and probably the ground substance. Macrophages become very active in regions of inflammation and are known to engulf foreign materials. Mast cells are phenomenal cellular elements, containing a vast array of biopotent chemicals, such as heparin, histamine, and immuno-logically related proteins. Since all organs (even the brain) are surrounded or penetrated by these cell types, connective tissue cells are in a strategic position to mediate diffuse effects over the entire body.

Body Mechanical Vibrations

All physical systems have essential natural frequencies or vibrational bands at which they are most elastic. In some such as glass or suspension bridges, maintained application of the natural frequency results in greater and greater vibrations until the object shatters or is destroyed. Biological systems do not shatter, but they must still dissipate energy acquired in this manner.

The human organism is a biophysical system with an intrinsic frequency due primarily to the skeletal muscle mass. Sensitive measurements indicate that the human body vibrates at frequencies between 6 Hz and 20 Hz. The amplitude of the vibrations, usually one to five micrometers (10^{-6} m), is invisible to the naked eye in most situations. When the muscles are tensed or during hyperthyroid conditions, the vibrational amplitudes can exceed 50 micrometers and become noticeable. Frequency and amplitude of the human body are influenced primarily by body temperature. Consequently, menstruation, fever, or hypothermia markedly alter body vibrations. After death, microvibrations of the body continue for more than an hour.

Although the vibrational profiles of the human musculature are very complicated and composed of hundreds of discernible averages, general patterns emerge. Theoretically, application of similar vibrational stimuli from the environment could enhance their amplitude and induce greater force or strain within the muscle mass. These changes would be diffuse and, again, detected by that vast intermediary system between knowing and detection: the autonomic nervous system. Candidate stimuli of the human vibrational system are the inaudible components of wind and infrasonic fields from moving air masses. Ultrasonic sources should not be excluded.

Steady State or Direct Current (DC) Electric Fields

Under normal conditions, a potential difference of about 10 microvolts (10^{-2} V) exists between the spinal cord and the hands or feet. The hands and feet are usually negative with respect to the spinal cord, while the front of the head is usually negative with respect to the back of the head. Several scientists have proposed the body polarities as a primitive detection system.

A human being about two meters high would behave like a large antenna that could summate or detect small changes in atmospheric electromagnetic events requiring minutes to hours to occur. Changes in geomagnetic disturbances or atmospheric electric field gradients could mediate their effects through this system. The most typical biological response would involve slow pain of the arthritic variety and general loss in neuronal potentials, thus inducing alterations in consciousness or general activity. That these changes may be sensitive to other aspects of the weather matrix has been implied. Ravitz (1953) found impressive relationships between lunar phase, (d.c.) potential fluctuations of the body, and agitation in psychiatric patients.

Time-varying Electromagnetic Fields

The human being can be described as a cluster of variable electromagnetic oscillators. The most intense oscillators, the heart and the brain, display electromagnetic fields with power peaks around one cycle per second (1 Hz) and 10 cycles per second (10 Hz), respectively. Although these fields cannot be measured more than 50 centimeters from their outside surfaces, these organs can be considered complicated clusters of electromagnetic flux lines.

Higher frequency electromagnetic fields, or harmonics, have been measured in the heart and brain as well as in other organs. Some of these bands are very narrow, while others show very wide ranges. Bursts of frequencies within the radiowave and microwave bands have been measured.

Certainly the most pervasive source of electromagnetic fields emitted from the body is the infrared band, the majority of wavelengths occurring around 10 micrometers (μm). These fields can be measured more than one meter from the body surface. Few empirical studies have analyzed this diffuse gradient system in detail.

The weather matrix and geophysical environment generate a vast range of electromagnetic fields whose frequencies, intensities, and shapes overlap conspicuously with those generated by the human being. Coupling and information transfer between biogenic fields and environmental fields are expected from crude physical experiments.

3

THE HUMAN ORGANISM AS A COMPLEX ENSEMBLE OF FEEDBACK SYSTEMS: EXPECTED PROPERTIES

With so many sites of possible interaction between weather stimuli and the body, a normal person might feel overwhelmed. No doubt, the human organism is complex but not beyond conception. We require a simple model that will allow a general grasp of the multitude of events occurring in the body. However, be forewarned. Models can be dangerous since they ignore individual differences and smooth out everyday biological variations. A model loses scientific usefulness when it becomes a faith instead of a tool.

SIMPLE FEEDBACK SYSTEMS

One model that has been used to describe living organisms is the feedback system, servomechanism, or homeostatic principle. Essentially, this model assumes that the components of a given system are interconnected so that any discrepancy between input and the intended output, as determined by the properties of the system, are minimized. Life support systems are totally dependent upon such types of reactions.

An elementary design or schematic of a feedback system involves a sensor, a comparator, and a compensator. The sensor responds to various sources of input while the compensator is functionally related to output. A simple diagram of a feedback system is shown in Figure 1. Note that the input is fed into the sensory apparatus, which in turn stimulates the comparator. The comparator contains a set point or predetermined value, influenced by either genetics, or learning, or both, with which incoming information is compared. If the values of the input are equal to or within the normal

35

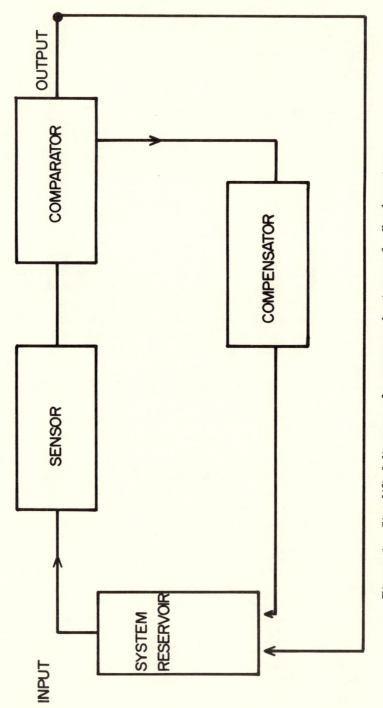

Figure 1 Simplified diagram of a servomechanism or feedback system.

36

variation range of the set point, the feedback system is not stimulated. Activity continues as usual.

However, if there is a discrepancy between the set point and the input values, information is sent from the comparator to the compensator. The compensator responds to reduce the discrepancy between the input values and the set point values. This output, called feedback response, is released into the total system (reservoir) and ultimately becomes part of the sensory input reaching the sensors. As the response from the compensator continues, the information flow in the system reaches the level of the set point. When this occurs, the comparator no longer stimulates the compensator and the feedback system is shut off.

A Mechanical Example

The most common example of a mechanical feedback system or servomechanism is the household thermostat for temperature control. The sensor is a thermometer, the comparator is the thermostat, and the compensator is the heating elements throughout the house. If the comparator has been positioned at 70°F, this temperature is the set point or predetermined level of the system. As long as the ambient or room temperature is maintained within that value, plus or minus a degree, the compensator will not be activated.

Someone opens a window in the room for a few minutes, and the temperature falls to 65°F. The change is detected by the thermometer, which communicates the value to the thermostat. Now there is a –5°F discrepancy between the set point and the ambient temperature or input. The circuits to the comparator are opened, and the heating elements become active. The temperature of the room slowly increases until the discrepancy between the set point and input values equals 0; the heating elements turn off automatically.

Biological Examples

The human body has been described as a complex of feedback systems that maintain essential life support operations. As in most other mammals, the human body is maintained at 37°C, plus or minus a degree, through a biological thermostat located primarily in the anterior region of the hypothalamus. The sensor portion appears to be special properties of the cellular membranes that exist in close contact with the local blood flow. The comparator is a class of temperature-sensitive metabolic reactions, which, when altered above or below a given set point, induce numerical patterns of action potentials. Compensatory responses follow.

If the temperature of the blood passing the sensor-comparator cells is too high relative to the set point, compensatory dilation of the peripheral blood vessels occurs reflexively. If the local temperature around the sensor drops below a critical level, the compensatory consequence is peripheral vasoconstriction. Other reactions will also be evoked, such as increased adrenalin output, increased blood flow, and recruitment of glucose (blood sugar). All are involved in the compensatory responses. Once the blood temperature passing the sensors falls within the range of the set point, the compensatory systems return to baseline.

Food consumption is controlled primarily by hypothalamic homeostats. The central unit of information in this system is the relative concentration of glucose in the blood passing hypothalamic nuclei. Sensors in the membranes of glucose-sensitive cells stimulate comparator reactions in the cell bodies. The reactions have not been specified but must be involved with glucose-dependent chemical reactions. Most likely, the glucose-sensitive cells contain quantitatively more reactants and products rather than qualitatively different elements.

If the glucose level of the blood passing near the "glucostat" cells falls below the criterion level, compensator reactions begin to occur. The animal becomes more active, sensations of hunger begin to predominate, and stomach contractions or hunger pangs are reported. The food-deprived animal then consumes available food, which results, after a brief period, in the elevation of blood glucose. Once the glucose level reaches the set point of the gluco-stats, various compensators are terminated.

In fact, if too much food has been consumed and the blood sugar has surpassed the set point levels, another type of compensator pattern occurs. The person may vomit, become lethargic, feel dizzy, or report apprehension. Psychologically, the overstuffed person may suddenly find the concept of eating completely aversive. He may swear to himself never to eat again, until the input falls below the set point. Then all aversive thoughts are forgotten.

Behavioral-Psychological Examples

The feedback model can be applied to behavioral or psychological phenomena as well. One popular example involves optimal amounts of perceptual-sensory stimulation. The sensor in this situation is the entire sensory field and perceptual transform functions of the body, while the comparator involves memory. At a physiological level, the primary correlate of the comparator is the background activity of the neurons.

Suppose the amount of information processed (the numbers of total action potentials and their patterns) per unit time has been determined by experience/genetics. There is an optimal amount of sensory input as determined by the set point. If the amount of information is greater than the set point, feedback mechanisms are activated. Although the details of the compensatory responses would vary as a function of the reward/punishment history of the person, a number of general sensory overload behaviors would become evident. The person would begin to ignore classes of social stimuli, which may have survival value, as if they did not occur. The person may spend more time by himself/herself, away from group contact or stimulation from synthetic media such as television or radio. At the extreme, when the sensory overload is too far above the set point, the person may show severe behavioral isolation and psychological withdrawal, typified by symptoms of autism, sensory blunting, or schizoid reactions.

When the amount of input is less than the set point, compensatory patterns occur in the opposite direction. The person spends more time in complex social situations, such as parties, movies, or bars. At home, he/she may listen to the television, radio, and record player simultaneously. The latter situation is quite typical when people who are habituated to a particular amount of social input are isolated during weekends. Telephone calls to friends become more frequent, and episodes of any mundane stimulation, from staring out the window to watching logs burn in a fireplace, suddenly become reinforcers.

At the extremes of understimulation, bizarre behaviors occur. However, within the context of a homeostatic model, they are quite predictable. If the person has no control over the amount of sensory stimulation or he is maintained within an environment in which stimulation is minimal, the nervous system may synthesize its own input. The person may fantasize more frequently or hallucinate during waking hours.

The homeostatic/feedback model can be used as a simple description, and as somewhat of an explanation, of complex behavioral-biological behaviors. However, it is only a model. When only words are used to describe the model and to predict consequences from the assumptions, the understanding must be realized in that context. Words are very general descriptors of complex things. Sometimes they imply more understanding or prediction than is actually possible.

Intrinsic Properties of Homeostats

Homeostatic/feedback systems demonstrate a number of intrinsic properties: internal oscillations, negative feedback actuation, sensitivity/stability tuning, and response latency. Internal oscillations are noise factors within the system. They are usually measured as periodic or random fluctuations in the sensors or input flow that do not initiate comparator functions. As long as the amplitude of these fluctuations or their rate of change per critical unit time does not exceed set point values, compensator functions remain at minimum, but not necessarily zero, values. The tolerance of oscillation is determined by the intrinsic variability in the set point mechanism. In biological systems, this can be quite sloppy.

The most characteristic property of feedback systems is their tendency to reduce the discrepancy between the input and the comparator set point. This tendency is called negative feedback. All biological systems that maintain their vitality work upon this principle. Thus when a sudden surge in input occurs, either below or above the set values, the compensator response reduces the deviation.

When the change in input is too extreme or when the system is defective, positive feedback can occur. Positive feedback has many synonyms, such as vicious cycle or exponential exacerbation. In operation, this phenomenon involves a continuous amplification or increase in the error difference between the input and the set point. Instead of compensations that decrease the discrepancy, the compensators, usually in a failure mode at the time, increase the error. Biological systems die unless an override function intervenes.

The most common example of death-precipitating positive feedback is when a person suddenly loses about two liters of blood. Since the remaining blood volume is not sufficient to maintain arterial pressure, the flow of blood to the heart muscle through the coronary vessels decreases. With fewer nutrients, the heart muscle weakens, further diminishing heart pumping, which further decreases the flow of blood into the coronary vessels, further decreasing the strength of the heart pumping, until the cycle reaches the terminal phase.

Positive feedback of this variety is rarely found in biological systems for obvious reasons. Usually, overriding mechanisms shut down the system or the enhanced instability reaches a saturation point beyond which the system will just not respond. Electrical activity before and during epileptic behavior is an example of the latter response. Under normal conditions, the millions of neurons in a given area of cerebral cortex show statistical summations of firing in the range of from 15 to 40 summations (peaks) per second. The amplitudes of these peaks are very low and just barely move above the average noise fluctuation.

Slowly, greater and greater numbers of neurons are recruited into a single firing pattern. This process is reflected by lowered frequencies and elevated amplitudes. At some point, the vicious cycle begins. More and more neurons are recruited in a cascading manner as the amplitude of the discharge becomes higher and higher and higher, ranging from 10 microvolts (μV) to 100 μV to 1,000 μV and beyond.

Then the discharge stops. The electrical activity from the epileptic portion of the brain suddenly flattens out, and the entire operation is terminated transiently. The depolarizing wave responsible for the epileptic attack has saturated the system. No more neurons can be recruited since the entire population has been depolarized. As a result, that part of the brain remains silent with only occasional activity. However, unlike the positive feedback example of the heart, the system does not die. Other portions of the brain take over until normal function is restored.

A third property of biological homeostatic/feedback systems is a balance between sensitivity and stability. A system usually trades off sensitivity for stability and vice versa. Systems that are very sensitive to changes in stimulus input are usually very unstable, while those that are very stable are usually not very sensitive to small changes in input.

Low stability (instability) is manifested in servosystems by periods of oscillation of output or compensatory function around the set point or baseline, as demonstrated in Figure 2. These oscillations are decreased in amplitude, sooner or later, usually in an exponential-like manner. Return to stability or noise oscillations occurs at an everincreasing rate over time. The time interval is relative and need not be specified; it can range from minutes to hours to days, depending upon the biosystem.

Very stable, low sensitivity systems do not display postchange oscillations. Following alterations in stimulus information, the unstable period is brief, perhaps only one overshoot and undershoot around a baseline, before background oscillations recur. Such highly dampened systems characterize the normal central nervous system, even at the level of the action potential.

A final property of feedback systems is response latency, the period of stimulus presentation required before responding occurs. Every system has some such critical lag time. Response latency is a very important characteristic of servosystems, especially those involved with long-term adjustments. Such systems usually require changes in stimulation for from several minutes to hours before comparator functions occur. Differential lags in response time have clear advantages; they prevent needless activation of long duration feedback systems due to brief or transient presentation of stimuli. Can you imagine what would happen to your body if every time you opened the refrigerator door your physiology shifted to a cold stress profile for two days?

Failures in Biological Systems

Within the context of the simple model used here, failures in biological feedback systems can occur at the sensor, set point, or compensator mechanisms. The more common central means of influencing the system is, of course, the comparator. Here direct intervention or alteration of the set point produces a major change in the entire system. Inputs that were previously received as erroneous may now become a part of the criterion.

Changes in set point can occur from experience, disease, or genetic background. Viruses or their consequences can influence the set point thermostat neurons of the hypothalamus, resulting in general elevation of set point levels. Even though the ordinary biochemical reactions of the body, especially those in the brain, do not operate optimally at fever temperatures, manipulation of the set point allows hyperthermic conditions to exist. Since the elevated blood temperature is compatible with the altered set point, no compensator response occurs.

If damage occurs to the comparator neurons, lethal changes in the set point can occur. Some theorists now suspect an older person with insufficient blood flow in the hypothalamic region may actually freeze to death in a warm house. In principle, death would occur in the following manner. The disease state gradually reduces the blood supply delivered to the comparator neurons, thus producing cell death or marked alteration in function. As a consequence, the set point is lowered gradually. The ordinary blood temperature of about 37°C or 98°F becomes erroneous, and compensator responses occur. The vasculature dilates to release the heat, and even a general decrease in heat production may occur.

As a result, body temperature begins to decrease. At lower temperatures, the muscles become stiff and the person has a hard time thinking. Even though core temperatures at the lower temperatures are adversely affecting physiological reactions, the thermostat neurons continue to exert influence. The temperature of the body continues to fall until it reaches the abnormal set point. At the extreme, the person dies by freezing, even at a room temperature of 80°F.

Learning can also influence the set point. Factors such as acclimatization or excessive/insufficient eating may alter the set points of thermostats and glucostats of the brain, respectively. Often the latter alteration is inadvertent. Concerned parents overfeed their young children, thus maintaining high glucose levels within the blood. As the glucostats mature, this level becomes the set point and all future adult inputs are compared with it. Even though the excessive glucose levels are not required and result in excessive food consumption and weight gain, the person is faced with a life of feeling hungry.

Other malfunctions in biological feedback systems can occur at the compensatory mode. Usually the net difference between the input and comparator set point and the amount/degree or qualitative change in the compensator output is specific. These values are usually well within range of normal, optimal adjustment times for the organism. However, if disease or learning history alters comparator function, a number of odd responses can occur.

For example, slight decreases in core blood temperature are associated with an almost linear decrease in peripheral blood flow and vessel diameter. In defective compensator systems, a slight decrease in temperature or stimulation of those neurons may result in an all or none type of vascular change. Suddenly, without warning, massive vasoconstriction occurs and the person becomes "white as a ghost" and reports anxiety, apprehension, and related foreboding sensations. If the stimulation continues, the person may have sustained hypothermia of the extremities, which may appear gray or even blue in color. In extreme cases, the excessive loss of blood flow due to the overconstriction can result in tissue death and gangrene.

Another type of compensator failure is related to excessive overshoot mistakes. As mentioned, the sympathetic and parasympathetic nervous systems work in a reciprocal antagonistic fashion. Stimulation of one is associated with suppression of the other, followed by a rebound influence from the latter. Suppose the person has been sympathetically stimulated. Large amounts of required glucose have been released into the blood flow to meet energy demands. If excessive compensatory overshoot occurs, this increase in glucose could soon be followed by an insulin surge, associated with parasympathetic overshoot. The person would show a contradictory hypoglycemia or loss of blood sugar due to the excessive usage of glucose.

Such breakdowns are not usually manifested during steady-state conditions, when the input is maintained within the noise level. There is no requirement for feedback components of the system to be active. Any deficit or instability within the "circuitry" will not be displayed. However, when changes in input do occur, the feedback systems are activated and any defect within the comparator or compensator system is unmasked. During the periods of adjustment to stimulus change, an otherwise normal system can suddenly show failure responses.

Relevance to Weather

The relevance of homeostatic/feedback models to biometeorology should be readily apparent. First, the hypothalamus and autonomic nervous system, the most common direct correlates of weather-related changes, are easily described in homeostatic terms. Actual predictions, both qualitative and quantitative, can be made once the parameters of each system are known. Secondly, weather is a source of input change. One would expect that any instability or problem within the various systems would be increased during adjustment periods to weather movements. These effects would exist as transients that may not be apparent if vulgar averages are taken of organismic output.

BASIC SIGNAL SHAPES AND PATTERNS OF RESPONSE

Intrinsic limitations and potentially abnormal deviations in individuals and populations become more apparent during periods of stimulus change. During such periods, the input stimuli can display a multitude of patterns. Three general stimulus shapes have been called step signals, ramp signals, and impulse signals. Each signal or input generates a particular pattern of occurrence in biological homeostats (Figure 2).

Step Signals

Sudden, abrupt changes in input are called step changes. Step changes are so sudden and intense that the comparators can neither accommodate nor anticipate the optimal compensator function. Consequently, as shown in Figure 2, a human homeostat becomes unstable after a step input. This instability is characterized by overshoots (overcompensation) and undershoots (undercompensation) of output until the discrepancy between the new input and set point approaches zero once again. The duration of the instability is related to the damping factors of the system.

During periods of poststep instability, the system does not respond systematically to changes in the environment. Sometimes changes in input will generate another burst of oscillation, depending

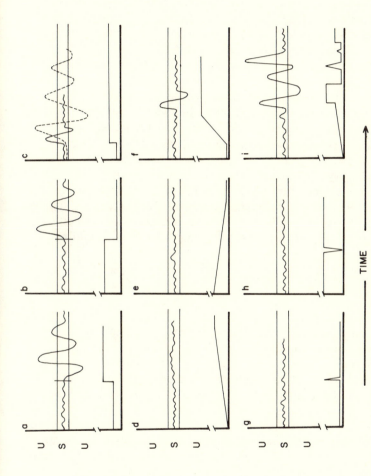

Figure 2   General response patterns of homeostatic systems to different types of input signals as a function of time. "S" indicates stability regions, while "I" indicates regions of instability. Sample input signals include step (a, b, and c), ramp (d and e), ramp-step (f), impulse (g and h), and a signal complex (i). A comparison of a high damped (continuous line) and a low damped (dotted line) system is shown in c.

upon when the stimulus is applied, while at other times the same stimulus may produce no observable effect. Odd interactions between otherwise ineffective weather stimuli and the unstable system can occur during extremes—peaks and troughs.

Poststep instability should not be treated lightly. At the total organismic level, sudden alterations in lifestyle, such as divorce, change in residence, new job, or even retirement, can be followed by higher risks of disease. It is not yet clear whether a step change in lifestyle increases susceptibility to immunological disease stimuli or whether the change unmasks or enhances deficiencies in the person's weakest organ system, such as the heart, connective tissue, or autonomic nervous system.

If weather stimuli primarily influence the autonomic nervous system and the behavior primarily correlated with it, sudden steplike changes in the weather matrix should produce perturbations in this system. Although personal conditioning histories primarily determine the actual details of behavior, general patterns would involve emotional instability and greater notice of such behavior. Otherwise normal, but latently unstable people would display enhanced bouts of depression, irritability, spontaneous aggressive outbursts, apprehension, general agitation, and water balance difficulties. Once adjusted to the step input, these people would return to normal as if nothing had occurred.

Ramp Signals

A ramp change involves a slower alteration in stimulus input to the comparator over time (see d and e in Figure 2). Since the rate of change is very slow, the comparator systems can respond quickly and accurately. Slow moving air masses and fronts that require several days to approach a human population are biometeorological examples of ramp inputs.

Ramp inputs have two important consequences. First, because the stimulus change can be accommodated by the body's homeostats, little or no instability occurs. Second, the absolute change in stimulus input over long time periods, such as days, can be very great. Whereas the same absolute change in input within a few hours would have been noticeable and disturbing, the same change over a longer period is barely noticeable.

Although ramp inputs are not traditionally associated with instability, a special condition of this signal may have biometeorological relevance. In systems that respond to rates of change primarily and to absolute change secondarily, very small ramp inputs could go undetected. If these inputs were just below the increment of detectability by the sensor, they would slowly summate to large deviations from the set point over time. However, at some point, the large deviation would be detected and the comparator would respond as if a step input had been presented.

This input pattern is shown in Figure 2f. An interesting example of this phenomenon can be seen in theaters in which dimming lights are used. If the decrease in illumination is slow enough, the audience continues to talk as if no stimulus change has occurred. However at some point a critical threshold is reached, and the audience notices the change. A marked alteration in general activity then occurs. If illumination had been suddenly dropped, the audience would have responded to a change much lower in total intensity than to the total decrease associated with the slow dimming.

Impulse Signals

Impulse shapes or signals are sudden but short-termed changes in input that are insufficient to induce major alterations in homeostatic outputs. Although intensity of the input is certainly important, the actual duration of exposure is the critical aspect of the impulse. As long as duration is less than the critical detection time, response latency (that is, delay in response time), or peak to peak time of the system's internal oscillations (background noise), no response should occur.

During impulse periods, the system can be exposed to intensities that would produce great instabilities under longer presentation periods. Since the stimulus is removed or disappears before the system can respond to or detect it, no instabilities occur. No doubt, the critical time period varies among systems and can involve nanoseconds, microseconds or even hours, depending on the system. At normal intensities, the neuron does not respond to a signal lasting a nanosecond, yet it does respond to signals within its normal information mode of a millisecond.

Compared with central nervous activity in general, the autonomic nervous system has a relatively longer latency of response to new stimuli. Before the adrenal glands release corticosteroids, the pituitary secretes adrenocorticotropic hormone (ACTH), blood sugar elevates, and general arousal occurs, the appropriate stimulus must be present for several minutes or be mediated after short presentations by ruminating thoughts. Less intense stimuli must be present for periods proportional to their amplitude. From tens of minutes to hours may be required before any general response occurs. If the stimulus is removed before that critical period, the system continues to respond as if nothing had occurred.

The latency of response in the autonomic nervous system allows the human being to be exposed to marked changes in the environment for short periods of time without massive changes in adjustment. Without this capacity, a short winter run between buildings would elicit cold adjustment responses that would continue for days. Merely opening the refrigerator door and exposure to the 50°F air for 10 seconds while looking for a midnight snack would throw the body into cold stress reactions.

A Note on Time Intervals

The concept of step, ramp, and impulse signals can be helpful in understanding how slow or fast approaching air masses might produce disturbances within human physiology and behavior. However, these signal shapes are relative to the temporal interval at which the system operates. To a system that operates and responds in intervals of days, an increase in external temperature of 1°C per hour would appear as a step-change since the system summates events into daily increments (24°C per day). On the other hand, a system operating at time intervals of an hour would respond to this change as a ramp. Similarly, a stimulus presented and withdrawn abruptly after one minute would occur as an impulse to a system using temporal increments of hours, but would appear as a step to a system using increments of milliseconds.

Consequently, it is critical to isolate the intrinsic temporal increments of the system in order to evaluate the shape of the input. Although it would be theoretically appealing if a stimulus had to exist for at least one temporal increment of the system's cycle before any effect would become apparent, detailed experiments have not been conducted. If this assumption were correct, an electrical stimulus must be presented for at least one millisecond (the increment of the electrical activity of the nervous system) before a response would occur. A weather stimulus would have to be present (changing) for at least one ortho/parasympathetic period: about 24 hours.

A SPECIAL CLASS OF SIGNALS

Theoretically, all electromagnetic and mechanical systems have natural frequencies at which they can be influenced with minimum applied energy. When natural frequencies are applied to these systems, energy is absorbed or added to the already low-level propensity to oscillate. These oscillations increase to amplitudes capable of modifying or destroying the entire system if the added energy is not removed. Classic examples of natural frequency effects can be seen in suspension bridges to which optimal low frequency vibrations are applied. Slowly the bridge begins to oscillate more and more until it disintegrates.

Very often the energy required to produce such drastic effects can be quite low and, if applied at some other frequency, would not be effective. Suspension bridges can be destroyed by foot soldiers marching in rhythm, whereas the same number of soldiers, walking in a random manner, would not be effective. Ordinary sound energy, when concentrated into a particular frequency compatible with that of crystal, shatters the crystal. The same relative intensities at different frequencies would not be influential.

Biological systems do not fall apart or burst when natural frequencies are applied since there are a multitude of mechanisms to dissipate applied energies—these systems are more elastic than solid mechanical structures—and since the intrinsic vibrations of the human body are so complex that continued application of the optimal vibration pattern for sufficient periods is highly unlikely. There are some theorists who insist that if the optimal complex (and it would be complex) of mechanical vibrations were ever applied to the human body in optimal amplitudes, the body would literally fall apart.

Parts of the human body can be driven by a multitude of external stimuli. One common example is alpha driving. The human brain displays a range of electrical (EEG) activity between 1 Hz and 30 Hz, which has been divided into various bands. The alpha frequency band has been defined as EEG activity between 8 Hz and 13 Hz and is usually associated with relaxation and imagery. Light flashes applied within this frequency range can drive the alpha rhythm of some human brains easily. Once the brain frequencies have locked onto the applied light flashes, brain electrical activity can be increased or decreased by changing the frequency of light flashes. If the applied light flashes drop to below 8 Hz, the prone person displays epileptic discharges.

Possible special signals at lower frequencies (or, more appropriately, periods) have not been studied or considered seriously. Are there environmental stimuli that show variations in the order of once per hour (that is, a period of 3,600 seconds or 0.0003 Hz) or, even more difficult to detect, once per day (that is, a period of 86,400 seconds or 0.00001 Hz) or even longer? Are there biological homeostats with intrinsic frequencies that would respond to these signals? What about the slow response systems, such as the immune complex and some aspects of endocrine systems, that display fluctuations in the order of weeks or months? Do they have special signals by which they would optimally be driven?

There are zeitgebers or time givers, such as light/dark and noise/silence cycles, that can modulate homeostatic outputs. Changes in the actual onset times of these periods, if done appropriately, can "drive" many blood and physiological reactions. Peaks in biological output can be shifted to occur in periods that had previously been troughs of inactivity. Traditionally, however, zeitgebers are not considered as special classes of signals.

FACTORS CONTROLLING THE INTENSITY OF WEATHER RESPONSES

The absolute intensity of responses to weather changes is not simply related to the intensity of alterations in the weather matrix. Several limiting factors influence the absolute and relative magnitudes that the compensator or response component of the human homeostats can display. These are initial values, compensatory overshoots, and response exhaustion.

The Law of Initial Values

Let half the length of a plastic ruler be suspended in midair over the edge of your desk. Now pull the end of the extended ruler up or down. Note that a slight deviation from the plane can be accomplished with little force. However, as you try to bend the ruler father and farther from the plane, more and more force is required. In order to obtain a ten-degree deviation that required only a little force at the plane, you now have to exert from five to ten times as much force. In other words, the stimulus input has to be between five and ten times greater in order to obtain the same absolute change in the response of the ruler.

Behavioral-biological systems work in a similar manner. Each system has an upper and lower limit of response. As the system is stimulated toward either of these boundaries, more and more external stimulation is required to elicit the same absolute change. Whereas a slight stimulus produced a remarkable effect when the system was operating between the boundaries, the same stimulus would have little effect once the system approached one of the boundaries. As the system is pushed near its ceiling or response asymptote, more and more stimulation is required to obtain the same deviation from baseline.

Heart rate is an example of this phenomenon. When the person is sitting down and resting, a human heart usually displays from 70 to 80 responses or beats per minute. To produce an increase to 120 beats per minute, very slight stimulation is required, such as walking or watching an exciting TV show. However, in order to increase the heart rate by another 40 beats per minute, to 160 beats per minute, vigorous exercise several times the intensity of the easy stimulation producing the first 40 beats per minute increase is required.

The law of initial values demands that we know the baseline of the system's output and how close this baseline is to the upper and lower limits of the system. Without this knowledge, prediction becomes very difficult and misleading. Suppose you have selected a group of people for study. You measure their baseline heart rates and find that they average 150 beats per minute. Since you have not determined the limits of the heart system, you do not realize that these people are already stimulated excessively. Now you apply a meteorological stimulus to the group and measure their heart rates: the average is 155 beats per minute. You conclude that the 155 beats per minute divided by 150 beats per minute is a trivial 3 percent increase.

Another experimenter selects a group whose heart rates average 80 beats per minute. He applies the same meteorological stimulus and finds, to your surprise, a significant increase in heart rate. His subjects show elevations of 40 beats per minute to an average of 120 beats per minute, a 50 percent increase. You ponder. How could the same stimulus produce such great effects on the other experimenter's subjects? Once again the problem of initial values has obscured a potentially important effect.

The excitation history of a system should be known before the presentation of a test stimulus. If a system is very close to the boundary of output, any further increase will be very small in absolute terms while any decrease will have a very large absolute range to cover. Thus a stimulus that drives near-boundary output to normal range will evoke a much larger absolute change than one that attempts to push the system further to its boundary.

Suppose a score of 30 on an anxiety test falls within the normal range of values, while a score of 60 suggests very high anxiety. You have two populations, one averaging 40 and the other averaging 60 on the scale. In order to reduce anxiety, you give the high anxiety group meditational therapy while the other group is left as control. You should shudder. No doubt this is a very sloppy but common design among promoters, fake therapies, and rip-off groups.

Caught in the enthusiasm of possibilities, the young experimenter ignores absolute changes and baselines and only looks at relative change. Using a popular scoring procedure, he calculates the relative change by subtracting new scores from old scores, dividing by new scores, and multiplying by 100 percent. The apex of his head tingles and he gasps. The group that received meditation showed 100 percent improvement while the control group showed only 33 percent improvement. What great effects!

However, the effect is illusionary. A quick calculation of the absolute change indicates the following. A 100 percent improvement in the meditation group indicates that its score dropped from 60 to 30 (that is, $60-30/30 \times 100$ percent = 100 percent, while the control group's dropped from 40 to 30 (that is, $40-30/30 \times 100$ percent = 33 percent). In actual fact, there is no difference between the meditational treatment and the control treatment. Both procedures produced scores within the normal range. The apparently great effect of the meditational treatment occurred because that group was already highly anxious.

In systems that display antagonistic behaviors, such as the autonomic nervous system, the absolute position of the baseline activity determines not only the net change but also the direction of change. If the system is already near maximum sympathetic activity, the presentation of a cold air mass, which is presumably more sympathetically effective, would produce less net further change than the presentation of a warm air mass. With the warm air mass, which is presumably parasympathetically effective, the net change would be much greater. Using diameter of blood vessels as a measure, the cold air mass presented to a blood vessel already sympathetically stimulated would produce little more constriction; the limit has been approached. On the other hand, a warm air mass will not only dilate the blood vessel to normal but may hyperdilate it to values typical of parasympathetic stimulation. In the latter case, the net change is very great, even though the external energy change is not impressive.

The relative position of baseline activity in the autonomic nervous system normally changes over a 24-hour period. During the day, the average person appears to be more sympathetically active, while during the sleeping hours, the parasympathetic nervous system predominates. From the general presentation of the initial values concept, one would expect that during the daytime, when the system is already sympathetically dominant, stimuli that initiate parasympathetic activity would produce a greater net change than stimuli that elicit sympathetic dominance. Similarly, during the sleeping period, stimuli that initiate sympathetic change would produce a greater net effect than parasympathetic stimuli. In both cases, the net effect is greater simply because there is more range of change in which to operate.

Compensatory Overshoots

The compensation principle states excessive stimulation of a system causes fatigue or exhaustion, followed by a compensatory overactivity from any antagonistic system. This principle is very applicable to the autonomic nervous system. Sudden failure in either the parasympathetic or sympathetic stimulation of an organ results in compensatory overcontrol by the nonfailing component.

A metaphorical demonstration of the phenomenon is easily created. Assume a ruler is maintained in midposition by two extended rubber bands pulling in opposite directions. When one rubber band is pulled, the ruler is displaced in the direction of the pull or stimulation. If one of the rubber bands breaks, the ruler does not return to midpoint. Instead, the ruler is pulled past midpoint in the other direction by the rubber band left intact.

Organs innervated by autonomic activity are prone to these sudden switchovers during periods of intense change in stimulation. Excessive sympathetic stimulation of the heart may push heart rate to some upper level until the system suddenly fails and the heart slows or even momentarily stops when parasympathetic stimulation occurs. Defecation is usually a parasympathetic activity. Under normal sympathetic stimulation, this function is not affected. During extremes in sympathetic activity, compensatory parasympathetic activity may occur in the intestines, and defecation occurs.

Such reversals may occur as a result of momentary exhaustion or depletion of transmitter substances from the terminals near the target organ. If the sympathetic transmitter substance noradrenalin is suddenly depleted, an organ receiving input from both autonomic components would be exposed to only acetylcholine, the transmitter substance of the parasympathetic nervous system. Consequently, the stimulus that initially produced the normally dominant sympathetic activity, but exhausted the transmitter substance, would elicit parasympathetic activity.

A similar phenomenon has been demonstrated within the parasympathetic nervous system. For example, if a physician injects acetylcholine, the typical response of peripheral vasodilation and a drop in blood pressure is observed. Technically, this response is called a muscarinic effect (after the effects of the drug muscarine). Now if these changes are abolished by injecting a parasympathetic antagonistic drug, such as atropine, an interesting reversal occurs. If acetylcholine is now injected for the second time, an opposite effect is observed: vasoconstriction and increased blood pressure (nicotinic effect). Since the first application stimulated one group of receptors that were inactivated by atropine, the second injection stimulated another group of receptors. Consequently, the same chemical produced the opposite result following a second application.

Compensatory overshoots or reversals, if they existed on a large scale in weather-sensitive people, would pose a major problem of data analysis. Depending upon the intensity of organ stimulation—and all major organs could be affected differentially—from a given weather matrix, the same weather matrix presented at a later date would not produce the same biobehavioral response. In fact, the response profile could be opposite to initial observations. A statistician, frustrated and psychologically overloaded by the great variability of individual responses, might conclude that weather never produces the same effect; consequently, weather does nothing at all.

Certainly, these reversal episodes would be influenced by drugs. Minor tranquilizers or stimulants given to the subject between the first and second presentation of the same weather matrix could modify autonomic tone. Whether pharmacological agents applied daily, such as caffeine from coffee, nicotine from cigarettes, tetrahydrocannabinol from marijuana, or ethanol from alcoholic drinks, would contribute to the variance remains to be investigated.

Response Exhaustion: Single Unit

Response exhaustion is a common source of variability in system output. This principle assumes a constant rate of production in response units. If a stimulus is presented that elicits these units, a critical period is required before the system is ready to display the same number of response units again. If the critical time has not elapsed, the presentation of the same stimulus will not produce any effect.

The principle is expressed in a variety of forms within different human phenomena. Immediately following depolarization (an action potential) of an axon, the stimulus that elicited the change is not effective. Immediately following ejaculation, the stimulus that produced erection is not sufficient to produce another erection. However, after a critical duration (in the order of milliseconds and minutes for the two systems, respectively), both systems are ready to display impulses once again.

Depending upon how frequently the initially effective stimulus were presented, conclusions of relative effectiveness could vary markedly. If an electrical stimulus were presented 1 million times per second to a neuron, the neuron would appear to respond to only one of the 1,000 stimuli presented during the one millisecond recovery period. If different sexual stimuli were presented to the postejaculatory male once every second, the male erection response would appear to be very weak, occurring once between every 100 to 1,000 stimulus presentations. In order for the system to respond systematically, the time between stimulus presentations must be greater than the critical recovery time of the system.

Preliminary data indicate that the recovery time for systems reactive to weather changes are much longer than suspected, perhaps existing in the intervals of from days to weeks. The same weather matrix presented at intervals greater than this period would be expected to elicit similar response profiles. However, during some seasons, such as spring and fall, when several front systems move across a human population within a one-week period, weather-related responses would be expected to become unreliable and perhaps even reversible.

Response Exhaustion: Populations

The response exhaustion concept can be applied to groups of humans as well. In this form, each individual within the group or set becomes a response probability of 0 or 1, that is he/she either responds or does not. One of the most common measures of weather effects is not variations within individuals but alterations in the numbers of individuals displaying some response. Common examples are greater numbers of suicides during a particular lunar phase, increased numbers of traffic accidents during sudden drops in barometric pressure, and accelerated psychiatric admission rates following geomagnetic storms. The effects are always small but evident when large populations are involved.

Response exhaustion within a group of individuals or within a large set of units would work in the following manner. Suppose during any one-week duration, one out of every 100 people would approach a probability of one that he or she would display unstable behavior following a weather change. With a population of, for example, 10 million, a total of 100,000 people would be expected to show this tendency. A frontal system passes across the population on the first day, eliciting massive increases in unstable behavior. Three days later, a very similar frontal system passes over the population. Now there is a conspicuous lack of peculiar behavior.

Since the first passage has elicited the unstable behaviors in the number of units that would show this tendency sometime during the week, the population of responsive units has been exhausted. The second passage, three days later, moves over a population that has been depleted of people with unstable probabilities. If the second frontal passage had occurred one week later, when the population had recuperated and another 1 percent was ready to display unstable behavior, this passage would have been associated with a sudden surge in unstable behavior as well.

Suppose a frontal system had not passed over the population. What would happen to the 100,000 people with unstable tendencies during that week? Conceptually, the behaviors would have been distributed statistically throughout the whole seven-day period due to a vast variety of individual stimuli, from job pressures to lack of sleep. Because the behaviors were distributed statistically over time, they would not be conspicuous.

The change in the weather matrix would be a precipitating stimulus. Since weather changes involve very large areas within which millions of people are immersed, all unstable units would be exposed to more or less the same weather stimuli at the same time. As a result, statistically more unstable behaviors would be displayed in a brief period and, consequently, become very conspicuous.

Whenever averages are used to demonstrate a local weather or astrogeophysical effect, the response exhaustion principle becomes evident. Frequently, experimenters report their data according to the deviation above or below the mean for some period. Numbers of births are well known to cluster above the mean around days of full moon. Incidents of agitated behavior are more common during periods of frontal passage. Heart failures are more common during or following cold air mass influxes during the winter.

However, after these peaks in behavior, there is a clear decrease in frequency below the average. The peaks in births around full moons are followed by a very apparent decrease in numbers of births shortly thereafter. Such decreases are expected, if not required, consequences of depleting any given population of ready response units in a very brief period. Time is required before the system is ready to fire again.

PERIODIC VARIATIONS IN BASELINE

Variations in organismic output and baseline activity induced by the weather matrix are superimposed not only upon individual differences but also upon periodic variations. These internal clocks or biological rhythms contribute in a complex, waxing and waning manner to the human response profile. Generally, these variations are discussed according to a period, that is, the elapsed time between two successive peaks in periodic activity. Period is the inverse of frequency. For example, a frequency of one Hz has a period of one second. A frequency of 0.001 Hz has a period of 1,000 seconds or about 16 minutes.

Ultrashort Periods

Ultrashort periodic variations are more conveniently described in terms of frequency than of period. Phenomena with these time properties usually fall well within the normal psychological intervals of detection, perception, and understanding. EEG changes (from 1 Hz to 40 Hz or periods of from 1.0 second to 0.02 seconds), heart rate, muscle vibration, and peristaltic movements of the intestines fall within this range.

Short Periods

Typical short periods range between 30 minutes and 2 hours. One intense short period, most pronounced during sleeping hours, is the dream cycle. Depth of consciousness, as defined by the EEG, cycles from minimum to maximum during these intervals. To some extent, short periods are evident in waking activity. The upper limits of adult attention spans, fatigue intervals, and general response to repeated and continuous presentation of the same stimulus ranges from 90 minutes to 2 hours. There is even evidence that recall of newly learned material follows this pattern.

The Circadian Clocks

Certainly, the most powerful periodic variations are circadian or about 24-hour changes. All major response systems in the human body display some form of circadian variation typified by one peak and one trough in activity during a 24-hour period. Most theorists argue that the primary evolutionary source of this intense variation is coupled with the 24-hour rotational characteristics of the earth. The most common variant is light intensity.

Although light/dark cycles are important zeitgebers for influencing the daily peaks and troughs of animal activity in general, the human being is influenced less by this stimulus. Social and cultural factors are more important controls. However, the circadian cycle is still present under normal conditions. Even under abnormal conditions, such as social isolation or long-term containment (in a cave), variations in 24-hour changes are quite small. The major change is a lagging effect, whereby the peak to peak period becomes slightly longer. Over many days or weeks, this slight extension slowly rotates the beginning time of each cycle. Cycle onsets can be shifted by 12 or more hours.

Long Periods

The two most common long periods are weekly and monthly cycles. Since there is no a priori mechanism for the seven-day cycle, these variations have been frequently explained in terms of cultural effects, such as the seven-day week and work cycle. Although pleasing theoretically, few studies have measured people not bound by the seven-day ritual.

The approximately 28-day variation is manifested in both smooth and rough forms within a number of biobehavioral responses. Smooth occurrence is characterized by slow waxing and waning changes equally spaced over a month. Rough occurrences are characterized by abrupt peaks in response numbers that occur for a few days monthly, followed by little or no activity during the majority of remaining days.

Lunar changes have been a popular explanation for 28-day variations in menstrual cycles, birth rates, homocides, suicides, and general aggressive behaviors, such as rapes and assaults. However, there are many confounding factors both within the lunar explanation and in addition to it. For example, the rotation period of the sun is also about 28 days. Considering the well known solar-coupled variations in geomagnetic activity and flare frequencies, quick explanations of 28-day cycles in terms of the obvious, such as moon phase, can produce a special kind of lunacy.

Objectively, there is little doubt that 28-day periods exist within earth animal activity in general, from the sexual maturation of the Palolo worm in the south Pacific to alterations in human female hormone levels. The major question is not the phenomena, but the mechanism involved. Whether the effects are mediated by expectancy learned from old myths and monster movies, by indirect effects from weather changes or by direct effects from the possible sources has been the primary issue.

Seasonal Clocks

Three- to four-month rhythms have been noted in several animals. In cats, these periods occur in thyroid activity, body weight, adrenal secretion, and glucocorticoid secretion. People display similar periods for catatonia (a type of schizophrenia characterized grossly by prolonged motor immobility), thyroid activity, peptic ulceration, epileptic bouts, and food intake.

Annual Clocks

Smooth or rough annual periods occur in many normal and diseased conditions. Suicide rates, birth defects, aggressive behaviors, and a multitude of contagious diseases show seasonal variation. Behavioral fluctuations, especially depression and related disorders, contain a clear annual component. Whether these fluctuations reflect indigenous characteristics of the human species, seasonal variation, or cultural factors has not been determined.

Infralong Periods

The longer the period, the greater the problem of determining its existence and effect. No doubt, there may be periods in human behavior hundreds of years in duration. Demonstration of their existence becomes difficult when it parallels historical epochs.

Some apparent periods do reappear from time to time in biological studies. A great many of these cycles were isolated by E. R. Dewey (1970) and his group. Variations in human social behavior, from economic production to war, have been found within the following periods: 4 years, from 11 to 12 years, and from 22 to 24 years. Although the sun has been a popular culprit for the recurrence of these cycles in human behavior, much, much more data analyses are required before any certainty can be fixed to such speculation.

GENERAL CONCEPTS OF WEATHER STIMULI AND SYSTEMS THEORY

Michael D. Lebowitz (1973a) has developed similar approaches to the relationship between weather stimuli and human responses.

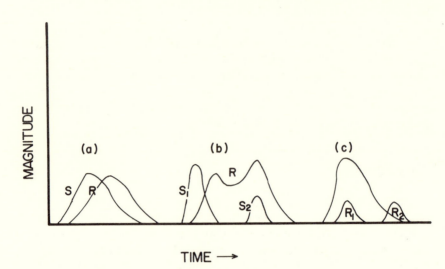

Figure 3   Some temporal and magnitude relationships between
            different stimulus (S) forms and elicited response (R)
            patterns (after Lebowitz 1973a).

Figure 3 demonstrates some of the qualitative response curves associated with the stimulus-response (S-R) concept applied to weather
stimuli. These three diagrams demonstrate, in essence, the problems of stimulus duration, response duration, lag time or latency
of response, and interstimulus intervals.

Figure 3a is a simplistic form of weather response patterns.
The stimulus arises and is followed by a response with a similar
pattern, the peak for which is delayed. Figure 3b shows the contribution of two closely adjacent weather stimuli and the complex form
of the response curve. Presumably, if the second stimulus had not
been presented, the response curve would have attenuated. However, stimulus 2 adds to the intensity of the response.

Figure 3c is an example of a stimulus of slightly longer duration eliciting two separate response curves. The two response
curves could be qualitatively similar or different. The first curve
could typify the most susceptible members of the population, while
the second curve might indicate the less susceptible members who
respond only to continued presentation of the stimulus. The second
option might involve two different types of disease responses alto-
gether, reflecting some unclear hierarchy of which physiological
systems fail first.

Lebowitz found that abnormal days are usually contiguous in time. A majority (60 percent) of abnormal days, those of high or low temperatures, were within four days of each other. The remaining abnormal days were more than eight days apart. Days of similar types of abnormal environments tended to be clustered from 74 percent to 90 percent of the time.

With this data, Lebowitz suggests that one should discuss a weather stimulus in terms of blocks of adjoining days. Single days or even pairs of days, from this point of view, might actually mask or exclude the interval of time (the block) that actually represents the weather stimulus as a biological event. In systems' theory, a block of days would be described as a step stimulus.

The definition of a response is slightly more complicated when one switches from words to numbers. Lebowitz notes, as do many other authors, that the intensity of the weather effect, using daily intervals, decreases significantly when data are lagged. Usually the highest correlations were found between the day of the environmental event and the day of the mortality measure. In some instances, the effects were maintained when the mortality measures were correlated with weather conditions of the previous one to three days.

Consequently, he defines a response as occurring more or less within the time frame of the stimulus. A response would occur within a day or two at most of the onset of the weather or the abnormal environment. Lebowitz contends that, for his mortality data at least, the mortality response occurs almost always on the day or the day after the start of the environmental stimuli.

Such models are important for isolating testable concepts in biometeorology. One can quickly generate a number of critical variables that should, intuitively or theoretically, influence the magnitude of a population response. These include the duration of the weather stimulus, the interstimulus interval (ISI), the magnitude of the stimulus, and the various interactions among them all.

4

LANGUAGE AND LEARNING
AND THE HUMAN LOCUS

Since more than half the data pool of classical biometeorology
is derived from self-evaluations, this type of measure must be given
a very close examination. If human behavior is a primary measure
of putative changes caused by the weather matrix, one must ask the
essential question, "What are the limitations and boundaries of this
measurement procedure?"

Human experience is not a simple phenomenon. It is influenced
by the language used to describe events as well as by the complex,
behavioral history of the human being. The person is more than a
simple homeostat responding to an ever constant set point. Due to
learning, human experience can change markedly. The set points
can be altered radically.

LEARNING AND CONDITIONING

A simple definition of learning or conditioning is the more or
less permanent changes in the behavior of an organism due to experience. Short-term and temporary changes in an animal's behavior,
such as sexual appetite or hunger, are not considered learning.
These changes are not permanent, but fluctuate as a function of
periodic processes within the body locus.

The essential theme in learning theory is the principle of
space-time contiguity. In the simplest form, it states that two events
occurring within an optimal period and space dimension will be related
by the body locus. The relation is due to some organic change within
the body volume that has been called memory or learning. Whatever
the organic bases, learning indicates that the two events become
systematically related.

Learning occurs even at the most fundamental level of the nervous system. For example, if one presents a light to the human eye, an electrical response will be evoked in the visual portion of the cerebral cortex. The relationship between the light presentation and the electrical change in this part of the brain is a genetic factor with which the person was born.

Now suppose a tone is simultaneously presented with the light. As you would suspect, the presentation of the tone would not elicit an electrical change in the visual cortex. Instead, it should evoke electrical alterations in the auditory part of the brain. This prediction would be confirmed.

However, after the tone and light have been presented together for a criterion number of trials (between 1 and 100, depending upon the situation), the effects of contiguity or learning become evident. If just the tone is presented to the person after this conditioning, a distinct response similar to that produced by the light will be evoked in the visual cortex as well as the usual response in the auditory cortex.

Research indicates that all known animals can be conditioned. Mammals, reptiles, and amoebas can be conditioned. Stated another way, even protozoan behavior can be modified as a consequence of experience. It should not be surprising that unborn animals can learn as well. Unhatched ducklings and even the eight-month old human fetus can be conditioned.

Two essential types of learning or conditioning models have been developed to describe modifications in behavior as a consequence of experience. These have been called respondent (or Pavlovian) procedures and operant (or Skinnerian) paradigms. Both assume space-time contiguity and the often arbitrary designation of a stimulus and a response.

A stimulus is any environmental event that can be demonstrated systematically to be associated, or correlated when mathematical analyses are involved, with the occurrence of a response. The duration of an experimental stimulus normally varies between a few milliseconds (the increment of time at which the neuron optimally operates) and several minutes. However, the upper and lower limits of effective stimulus durations are not clear. Depending upon the biological systems, some environmental stimuli may occur in increments of from days to months.

A response is defined as an observable event associated with body space. Most learning research has been involved with responses linked to skeletal (striated) muscle changes, such as lever presses, walking, or talking. However, any series of events within the body can be designated as responses. As long as these events can be pointed out, enumerated, and shown to other experimenters (inter-

observer agreement), they can be studied within the learning paradigms. Within these assumptions, electrical responses, smooth
muscle changes, blood chemistry alterations, and glandular weight
modifications within the body could be studied as responses.

The arbitrary nature of stimulus and response can be seen
most readily within the body locus. Whereas one event may be considered a stimulus in one series of events, the same or similar
event may be considered the response in another series of events.
Stimulus response chains occur throughout the body, overlapping
various levels of measurement and discourse. Consequently, a
stimulus or response event should be evaluated according to its
temporal characteristic or serial order, place within the body, and
the context in which it occurs.

Respondent Conditioning

The procedure essential to respondent conditioning involves
simple temporal contiguity: the pairing of two stimuli. To avoid
possible contaminations from genetic or congenital responses, one
must first determine whether a stimulus evokes a nonconditioned
response. The second stimulus must be neutral, that is, it should
not evoke the nonconditioned response.

In classic terminology, the nonlearned stimulus is called the
unconditioned stimulus (UCS) and the response it elicits is called the
unconditioned response (UCR). The association between the UCS
and UCR is usually genetically determined. In other words, a hot
stimulus will evoke the removal of the hand from the area. A rush
of air will elicit a quick closure of the eyelid. An injection of morphine will evoke sensations of euphoria and semicomatose experiences.

The neutral stimulus, before conditioning, should have little
or no effect on the UCR and must not evoke a response similar to
the UCR. After repeated presentations of the UCS and the neutral
stimulus together, the neutral stimulus is presented by itself. If
the stimulus evokes a response very similar to the UCR, the neutral
stimulus is called a conditioned stimulus (CS) and the response it
evokes is called a conditioned response (CR). Without the pairing,
the neutral stimulus would not have become a CS capable of eliciting
the CR.

The difference between the UCR and CR may be difficult to
quantify, although in both theory and principle the CR should always
be less intense than the UCR. Depending upon the measurement,
the CR usually has less amplitude, lower probability of display, or

more limited event options than the UCR. The distinction between the two is important when one is attempting to isolate whether a response has or has not been learned.

The number of UCS-neutral stimulus pairings required before the neutral stimulus becomes a CS varies in different parts of the body. As many as 1,000 presentations of two stimuli may be required in the electrical systems of the brain before conditioning becomes evident. If the UCS is very noxious to the organism, as defined by the organism's avoidance or escape behavior, very few pairings may be required. Conditioned taste aversion, whereby animals avoid a new food, can occur after it has been paired only once with a sickening episode.

If the CS is presented in several trials without the UCS, the dependent nature of the stimulus can be seen. As more and more CS presentations are given without the UCS, the CR becomes weaker or less frequent. At some point, the CS returns to a neutral stimulus since presentation does not evoke the response. Extinction has occurred.

The time required for extinction to occur is a function of several conditions. Extinction will occur more frequently if the CS is presented repeatedly and the UCS is never presented again. However, if the UCS occurs variably (every so often) in time, the CS may require more presentations before extinction occurs.

Extinction is more likely to occur when the situation is very similar to the initial conditioning characteristics. Although the primary aspect of extinction is the presentation of the CS without the UCS, secondary factors within the learning field contribute to conditioning/extinction probabilities. Sometimes an actual return to the learning situation is impossible, especially if the conditioning occurred during infancy or childhood. The conditioned organism has changed psychologically and physiologically. Extinction, in this condition, would be more difficult.

As should be surmised, respondent conditioning does not require voluntary assumptions for human behavior. In fact, the human being, like other organisms, is a passive locus with this type of learning. The behaviors displayed before or after the presentation of the stimulus may have little effect on the presentation of the UCS.

Consequently, respondent conditioning appears to be a primary mode of learning during childhood and especially early infancy, when the person is a passive, semihelpless organism who is inundated with an endless train of external and internal UCSs. Respondent conditioning is also a primary mode for those systems not commonly correlated with thought or awareness, such as the smooth muscle and glandular and vascular systems.

Since awareness is not a required condition for respondent conditioning to occur, especially for noncorrelated body loci such as body chemistry and perhaps even cell division, the adult human being can be conditioned hundreds of times without knowing it. Several psychosomatic diseases, such as ulcers, allergic responses, and hypertension, can be a consequence of this type of conditioning.

Respondent conditioning has been demonstrated in many response <u>systems</u> (response sequences that are highly intercorrelated) within the body. Several behavioral examples can be given. If milk (UCS) elicits satiation or terminates crying in a young infant, the milk can be called a UCS and the termination of the behavior a UCR. If a stimulus, such as the mother's facial features, is paired with the milk presentation, the mother's facial cues or presence becomes a CS that will also elicit termination of crying and satiation (CR). However, the CR elicited by the mother's presence is not as intense as the UCR elicited by the milk.

A more physiological example involves the injection of chemicals into the blood. If the intravenous injection of morphine (UCS) is associated with changes in brain chemistry (UCR), any coincidental stimulus could become a CS. The most common coincidental stimulus is the injection or the procedures just before injection. After conditioning occurs, the injection procedure without the morphine can produce similar chemical changes and subjective experiences.

In both the mother-milk and injection morphine cases, however, repeated presentation of the CS without the UCS will result in a gradual extinction of the CS potency. With each presentation, the CR becomes smaller and smaller until the CS no longer elicits substantial changes. Even at its most effective state, the CS never elicits a change as intense as the UCR associated with the UCS.

Respondent conditioning, for the most part, requires fewer trials when aversive stimuli are involved. Classic aversive stimuli are electric shock, noxious drugs, and direct or indirect stimulation of the punishment areas of the brain. However, any stimulus can be aversive to some organisms, even stimuli that may appear to be good or pleasurable to others. The aversive characteristic of a stimulus is defined by its effect on the response in question.

One daily example of an aversive UCS effect can be seen in those people who use a loud noise, such as a clock radio, to wake up every morning. A loud noise is usually an aversive stimulus to sleeping humans as defined by their behavior; it is removed quickly. In this situation, respondently, the loud noise is a negative UCS that elicits the UCR: suddenly waking up.

If a nearly inaudible click occurs a few seconds before the loud noise is presented, this click can become a CS after only one or two trials. The subject soon finds that she suddenly awakens (CR) a few seconds before the loud noise is presented. In some instances, the person may actually wake up and remember the click.

One behavior frequently displayed by children involves the following conditioning history. A painful stimulus, such as a slap (UCS), can elicit crying behaviors and uncomfortable subjective responses (UCRs). The elicitation of these behaviors by the slap is a consequence of the genetic makeup of the human body.

If the father's presence, for example, were paired with the presentation of the aversive UCS-UCR pattern, the mere presence of the father—without any slapping stimuli—could elicit CRs. In this instance, the CRs would include behaviors such as whimpering and unpleasant private responses, mostly of an anticipatory nature.

The larger the percentage of trials that the father's presence is associated with the delivery of pain, the more effective the acquisition of the CRs. If, for example, the father is home infrequently (once a month) and 50 percent of the time when he is home, his presence is associated with aversive stimuli, the CRs are very powerful and clear in direction. Most fathers are associated with both positive and negative stimuli. As a result, responses are more ambivalent.

Physiological responses can be conditioned to aversive stimuli as well. If rats are given an electric shock (UCS), a normal response (UCR) will be eosinopenia (a decrease in the number of a type of white blood cell) shortly thereafter. The presentation of the shock chamber without the shock does not elicit the change. However, once the chamber has been associated with electric shock, just placing the animals in the chamber can elicit conditioned eosinopenia. The chamber has become a CS that elicits the conditioned white blood cell change (CR).

Just about any stimulus can be paired with an aversive event. Such pairings, especially accidental associations between unrelated stimuli and aversive events, are common sources of phobias, irrational fears, and general superstitious behaviors. Since aversive stimuli can engage intense emotional sequences, the capacity to discriminate between complex perceptual relationships deteriorates. For example, a single episode of ridicule at a party may be sufficient to produce avoidance of and fear toward public appearance of any kind. In this situation, just being in public or, in more severe cases, thinking about attending a party, induces emotional behaviors.

Behavioral patterns emitted or displayed in the time interval between the presentation of a CS that has been paired with the later presentation of an aversive stimulus and the presentation of that stimulus have been called conditioned emotional responses or plain

anxiety. The human subjective experiences of anxiety range from
an unpleasant, vague, or tip-of-the-tongue sensation to deep feelings
of foreboding or doom.

Correlates of anxiety are loaded by contributions from the
autonomic nervous system and its interfaces. A low-level sympathetic
dominance is associated with peripheral vasoconstriction, for example, cold fingers and toes. More intense stimulation can induce
sweating in the palms for no apparent reason and variable vasculature
tone associated with psychological images of "butterflies in the
stomach." Heart rate can increase, and the subject may report
muscle tension. At more severe levels, the person may feel that
he is about to die or that something terrible is about to happen.

These behaviors represent the varied displays of a more
general principle: response disruption. Most, if not all, response
patterns are disrupted or even suppressed in the interval between
the presentation of an aversive CS and the actual occurrence of the
aversive UCS. People report not only anxious moods but also difficulties in continuing the behaviors, including thinking, that were
being displayed before the CS was presented. Sometimes the person
may not be aware of the CS presentation.

Anxiety has devastating consequences. Unchecked, anxiety
and its correlates can incapacitate even the most robust human being.
Psychologists have conceptualized a number of processes—ego defense
mechanisms—that the human being uses to keep anxiety out of awareness. Through procedures such as blaming other people, denying
that bad events can occur, or projecting one's own undesirable
behaviors onto other people, a person can reduce the anxiety associated with these events.

Intense anxiety is encountered infrequently in biometeorological
phenomena; the exceptions may involve those patients who respond
to some part of the weather matrix with intense aversive experience,
such as the pain of arthritis. In principle, however, anxiety paradigms are important to biometeorology. If any part of the weather
matrix affects the human population as an aversive UCS, any persistent stimulus occurring before or with that UCS could in time become
a CS or cue for the onset of low-level anxiety.

Although the person would not be incapacitated by such a stimulus, he would feel uneasy or just "off the day." At more intense
levels, weather stimuli that have become CSs may evoke feelings
of foreboding or a general disruption of ongoing behaviors. The
person may have a more difficult time thinking, especially for heavy
concentration tasks, such as mathematics, creative writing, or
problem solving. Speed and power conditions, such as making a
complex traffic decision within one second, would be disrupted
markedly. The source of the problem would be unclear.

Operant Procedures

Some behaviors are acquired because of their consequences, that is, association with the later presentation of an aversive or a positive stimulus. These behaviors are considered operant responses or instrumental responses since the delivery of the stimulus is dependent upon the animal's behavior. Voluntary behaviors are frequently correlated with operant learning.

In operant procedures, the probability that the response will be emitted again is a function of the stimulus following it. If the response is followed by an aversive stimulus (punishment), the response is less likely to occur again. If the response is followed by a positive stimulus (reward), it is more likely to recur.

The nature of the stimulus, whether reward or punishment, is defined not by philosophical assumptions about good or bad, but according to its affect upon behavior. In some instances, traditionally negative stimuli may become positive reinforcers if their occurrence after a response increases the probability that the response will reoccur. For example, electric shock paired with food can become a positive stimulus. Animals will respond again and again when the shock follows food, but will not respond if shock does not follow.

Two other types of response consequences can be logically deduced from the dichotomous division of response occurrence (increase versus decrease) and the dichotomous division of stimulus (negative or positive). The first is obvious. For example, if a response is followed by the removal of a positive stimulus, the response probability decreases. On the other hand, if a response is followed by the removal of an aversive stimulus, the response probability increases.

A number of everyday examples can be given for each type. If your smile is a positive stimulus, the display of your smile following someone's saying good morning to you would increase the probability that they will greet you again. On the other hand, if your smile is a negative stimulus, its presentation after the person's salutation would decrease the likelihood of another greeting.

Responses that are followed by the removal of an aversive stimulus are quickly acquired and maintained. If paying the bills on time is followed by the removal of an aversive, late penalty, bills will be paid regularly. If screaming at one's husband stops his arriving home late for dinner (the removal of an aversive stimulus), screaming responses will occur more frequently.

Many addictions to medication or quasimedication are due to this process. If taking a pill is followed by the removal of an aversive condition, pill consuming is increased. The aversive condition may involve general malaise, frank pain, or even tenuous conditions of psychological depression. If the pill, regardless of its actual chemical effects, removes the aversive stimulus, pill taking is reinforced.

Anxiety is a very painful or aversive condition that is described as incapacitating and even painful. Responses that are followed by the removal of anxiety will be reinforced. If a person is anxious about attending a party and drinking alcoholic beverages reduces the anxiety, the probability of drinking at the party is increased.

Unfortunately, responses that reduce anxiety are likely to be displayed in situations other than those in which they are acquired (generalization). Since anxiety conditions are so diffuse and common to some people that they occur almost anywhere and anytime, anxiety-reducing responses can develop an obsessive-compulsive nature.

Once acquired, some responses are very difficult to eliminate. The likelihood of extinction and the time required for complete extinction is a function of the pattern of reinforcement that has maintained the response. Rewarding stimuli or the removal of aversive stimuli that occur every so often will be maintained for long periods. This pattern of reward or punishment has been called a variable interval schedule.

Variable interval schedules shape up and maintain behaviors that are very resistant to extinction. Gambling, for example, can be very difficult to quit. Although the person gambles frequently, reinforcement (winnings) occurs variably in time. The number of bets, per se, may not determine winning probability. If chance alone is operating and the person wins "big" about every 7 days (range of from less than 1 to 14 days), gambling can be maintained for years. Long runs of bad luck may not extinguish the response.

Explanations, especially erroneous ones, can be maintained in this manner as well. Suppose the person believes that gray days make him depressed. And suppose the person is likely to be depressed about once every week: usually, a Sunday night or Monday. If the probability of a cloudy day is one out of four (more during the winter in some places), cloudiness and depression will be coincident about one out of four times; the overlap will be variable in time. Since the coincidence of the depression and the cloudy day would be tantamount to a reward situation, the likelihood that the person would conclude that cloudy days produce depression is increased.

Superstitious Conditioning

Responses that are displayed in close temporal proximity with the removal of an aversive stimulus or the presentation of a positive stimulus will be rewarded as well. Although, technically, these responses may not even be related functionally or causally to the removal of the aversive stimulus, they become strengthened due to closeness in time. A response that has been reinforced in this manner is considered a consequence of superstitious conditioning.

Usually superstitious responses are conditioned due to a variety of chance factors, although there are definite classes of behaviors prone to be conditioned in this manner. For example, members of a human culture do not usually resort to extreme or bizarre behaviors unless they are exposed to great deprivation. For instance, as the degree of drought increases in a particular region, the likelihood of increased frustrative behaviors or desperation responses increases as well.

However, the probability that it will rain, by chance alone, also increases the longer the drought persists. If the people do not die of thirst and hence do not report any further behaviors, the probability is quite high that they will display the extreme behavior— raindancing or some other form of increased ambulation typical of food/water deprivation—at the next rain. The occurrence of the rain reinforces the immediately previous behavior (raindancing) even though it was not technically associated with the delivery of the rain.

A more acute example is associated with the normal temporal decrease in pain intensity following stimulation. Suppose you severely bang your finger with a hammer. The intensity of the pain peaks within about a second but then decreases relatively quickly to a low-level throb within about 20 seconds. This sequence of events would occur regardless of how you behave.

However, suppose you curse intensely immediately after the hammer hits the finger. If the chain of cuss words is long enough (15 seconds), the pain will have been significantly reduced in intensity by the time the sequence is finished. Since the sequence was followed by pain reduction, it is reinforced even though it did not decrease the pain.

The human being's tendency toward superstitious conditioning is troublesome from an analytic point of view, since this type of learning can also occur between response systems. If a series of responses occur simultaneously with another series of responses that are actually associated with the removal of an aversive stimulus, both response sequences are reinforced.

If a human being is drowning, two behaviors have a high probability of occurrence: swimming and various thoughts about getting saved. The thoughts usually involve a parent symbol (God) since the drowning situation is very similar to the helpless behaviors of infancy and childhood. If the person is saved, both behaviors—

swimming and yelling for God—are reinforced, even though swimming technically prevented the drowning.

Superstitious conditioning can confound and distort the actual series of events when it is used as an explanation. In the previous example, the person would probably conclude that God as well as swimming had saved him. The culture who raindanced would probably report that the frenzied dance produced the rain.

Explanations are important removers of everyday anxieties. The most common source of everyday anxieties is events associated with uncertainty, unexpectancy, or the unknown. People may feel odd, unusual, or even about to die if they do not have an explanation for their ailments, especially if the experiences are novel. However, once an explanation is given—even though it may be technically worthless—the patient is given both a structure for the removal of anxiety and a response to display again if the condition recurs.

If the book in front of you suddenly became elevated and suspended in space one meter above your desk, you would probably feel a little anxious; the actual form may vary from a type of awe to stark terror, depending upon how you have learned to respond to new, unexpected events. However, if one were told the effect was due to hypnotism, the average person would report an immediate relief as well as some verbal statement such as, "Oh, that's how it happened!" In actual fact, the label "hypnotism" has given no information regarding the mechanism involved.

Catchall explanations for unexpected or odd events, including those occurring within the body, are usually selected from general concepts. Fate, God, bad luck, and weather are frequent choices since they are so vague and nonspecifiable that they cannot be easily tested and rejected. If they were specific stimuli and easily testable— directly or indirectly—the person would soon learn that the explanation is inappropriate. The anxiety associated with unexpectancy or novelty would return.

Once a phrase is used as an explanation and the anxiety is reduced, the person is more likely to use the explanatory phrase again. Since the phrase acquires reinforcing characteristics due to its association with anxiety reduction, it becomes more powerful with each usage and subsequent anxiety reduction. Soon, in many instances, just statement of the explanation is sufficient to eliminate the anxiety associated with the uncertain condition.

Unconscious Conditioning

One of the prominent illusions of the human condition is that each person controls all of his or her own behaviors. Through a series of relatively complex reinforcement schedules, a person learns to explain his or her own thoughts, actions, and decisions in terms of an egocentric source. This propensity is not surprising in light of the physiological-physical construction of the organism. Rewards and punishments occurring on or within the body boundary have a much greater probability of long-term impact (for example, memory) than those presented to other people or at some distance.

Unconscious conditioning is an expected property of human behavior when it is described as a three-dimensional locus of responses comprised of different systems. Each system is associated with internally consistent (highly correlated) responses. These different systems and their associated instruments or procedures of measurement comprise the levels of discourse that compose the disciplines of science. One can measure the human body in terms of physical, chemical, biological, behavioral, or sociological responses.

No doubt, these disciplines are gross, arbitrary distinctions of the body locus. Without diverging into problems of reductionism (the reduction of one discipline into the phenomenon of another), one can state that different disciplines represent internally consistent modes of measuring the probabilities of events within the body locus. The reductionism conflict appears to be more of a word game than reality.

Each system or cluster of interrelated responses contains characteristic behaviors. The muscle system, for example, is associated with increased movement of the total body (ambulation, talking, and so on) and with a variety of related chemical changes. When the muscle system is not displayed, these behaviors do not occur or are reduced in frequency of occurrence.

The vascular system displays analogous, interrelated responses. If the heart is stimulated, changes in the size, diameter, or volume of blood vessels occur throughout the body. If the heart is not stimulated, vascular reactions remain at some baseline level. Such reactions are evident and expected aspects of biological systems.

However, the same description can be applied to human private behaviors, popularly called thoughts, ideas, or experience. Although the only independent instrument of measurement for these events is verbal behavior—what the person says, writes, or implies through body motion—some obvious crosscultural and species-specific behaviors appear to occur.

When the private behavior system is displayed, the person is likely to report or to show evidence of awareness or consciousness. During the stimulation of this system, the person will report a conscious interaction with events occurring outside and inside the body locus. The person will imply or state that she is a self or person or some other description of a unique entity. Since rewards or punishments are most effective when applied to the body boundary, this presumption is not surprising.

Such unavoidable learning history develops not only a concept of the body boundary or self-concept, but also leads to a tendency to explain the environment in terms of an egocentric reference. When the private behavior system is operating, the person explains his or her own behavior in terms of the self. For example, "I am reading this book because I want to . . ." Even irrational and illogical impulses are explained, after the fact, according to the self.

When the awareness system is not displayed, the three-dimensional locus does not suddenly stop. Indeed, contrary to many assumptions, the general activity of the human being continues. Although the private behavior system is not displayed, for example, during sleep, many other systems are still operating. These systems may even dominate the total output of the body locus. Whereas during the waking state private behaviors may have comprised a major component of the total behaviors displayed, when asleep other systems may predominate.

Each system shares probabilities of occurrence with other systems. Some systems are more interrelated than others. For example, release of adrenalin into the blood stream is highly correlated with changes in heart rate and blood chemistry. Changes in the electrical activity of the brain are highly correlated with alterations in private behaviors. Whereas thinking about breathing can alter the rate of breathing, thinking about heart rate does not typically influence heartbeat nor does thinking alter the activity of smooth muscle. Such correlations are frequently determined by genetics.

Two systems or parts of systems can be correlated by learning. A simple example of contiguity is the conditioned electrical changes that occur in the occipital cortex in response to an auditory stimulus after a visual and auditory stimulus have been paired together. Another example is the control of heart rate by thinking after the person has been given feedback with respect to the consequences of his concentration upon the heart rate.

For a person to be aware of a series of events, private behavior must be displayed at the time of the event. In other words, the private behavior system and the other system must be correlated in time. Once this is completed, the person can report awareness or voluntary control of the event. Without this correlation, the person may not be aware of a series of environmental events and actually misconstrues or misevaluates their source.

Recent evidence clearly indicates that systems of the body can be modified without awareness by the human being involved. If the awareness system is not involved, the person cannot display any of the behaviors associated with it. If the system affected is not correlated with the occurrence of awareness, the person would not know what had happened.

One classic study involved the following procedure. Subjects were taken into a particular room and given a painful electric shock. Blood samples were taken in order to measure the levels of hormones associated with the event. Following this procedure, the subjects were hypnotized and given the suggestion to forget the entire episode. Apparently, some subjects respond easily to these simple procedures (which, incidentally, do not require going into a trance to be effective). Some psychologists argue that the instruction "forget" is similar to the mechanisms of repression popularly discussed by personality theorists.

Several weeks later, these subjects were asked to participate in another experiment. According to the subjects' reports and their various nonvocal behaviors (for example, body language), the experimenters concluded that they had no awareness of what had happened in the previous meeting. Consequently, the subjects were reintroduced into the room now associated with the painful experience.

Some subjects felt uneasy, although they did not give specific reasons. Common descriptions of their experiences were similar to tip-of-the-tongue sensations. Other subjects felt mildly uncomfortable, with sensations of slight foreboding. However, none of the subjects reported remembering what had happened there. Except for the mild alterations in feelings or thoughts, they appeared normal.

Other systems of the subjects were not as quiet. Analyses indicated that the subjects had responded to the room at a blood chemical level. Even though consciously they appeared normal, their blood chemistry was elevated to levels that were even higher than those associated with the initial traumatic association.

The study is quite applicable to biometeorological phenomena. If vast changes in a person's biochemistry can occur following the presentation of a single stimulus due to learning, then one wonders about the wide range of potential variations that could occur due to presentations of natural stimuli. The person would not have to be aware that the changes were occurring. In fact, if the autonomic or endocrine system were involved, such as the adrenal gland, significant alterations in steroid levels within the blood may be associated with only a mild uneasiness.

More sophisticated conditioning can occur without awareness, as demonstrated by the excellent series of experiments by Hefferline and his colleagues (Hefferline, Bruns, & Camp, 1973; Hefferline & Perara, 1963). In these situations, subjects were placed in front of a reward dispenser. The subjects' fingers were attached to a special amplifier that detected small muscle twitches well below the threshold of conscious detection. Twitches from one of the fingers were selected for reinforcement, a procedure that would result in the increase of twitch occurrences.

The procedure was simple. Whenever the finger twitch occurred, the equipment delivered a reward to the subject. If the finger twitch did not occur, no reward was delivered. As with other types of overt responses, twitches became more frequent with more and more reinforcement. During the extinction trial, the finger twitch was no longer associated with the reward. The response frequency slowly decreased.

When the subjects were asked after the experiment what factors were responsible for the delivery of the reward, the answers were as varied as the subjects. Some subjects concluded that their thoughts had controlled the delivery of rewards, a possibility that many subjects found fascinating. Still others reported that some other response, such as the twitch of an eyelid, was responsible. In all cases, the subjects were sure of their explanations.

The Hefferline studies demonstrated not only the capacity of the human organism to be conditioned without awareness; they also clarified an important principle of biometeorology: the problem of superstitious conditioning. If some response occurs very close in time with a response that is reinforced, the first response will be rewarded as well. In the previous case, for example, the accidental eyeblink that was displayed at about the same time as the finger twitch, was reinforced as well. In fact, any response displayed very closely in time to a reinforced response can be strengthened.

APPLICATION TO BIOMETEOROLOGY

The ease by which the human locus can be conditioned may have great survival value. However, this propensity for learning is the primary source of variability when attempts are made to determine the essential set of variables. Different learning histories in which different weather stimuli are presented in different temporal orders may be the fundamental source of the gross variability and the extreme individual difference found in biometeorological data.

On the bases of experimental data, one would expect the respondent conditioning paradigm to be immediately applicable to weather conditioning. If there exists some stimulus that, due to the genetic makeup or predisposition of the subject, elicits a UCR, the accidental pairing of some other environmental stimulus in close temporal proximity with the UCS could result in conditioning. After a few pairings of the UCS and the initial neutral (biobehaviorally ineffective) stimulus, the stimulus could become a CS.

Once this occurred, the CS would elicit a CR that would be extremely similar, if not identical within the limits of measurement, to the UCR. If the CS were then presented by itself, in a laboratory situation or in a correlational study, the evoked organismic effects would be similar to that of the UCS. The experimenter might conclude prematurely that the CS was bioeffective and produced a change similar to that of the UCS. The experimenter might extrapolate erroneously and conclude that the stimulus, a CS due to conditioning, causes the organismic changes due to genetic reasons.

Now a second experimenter reads the first researcher's results and becomes excited. He attempts to replicate the results with a group of subjects from a different weather pattern and with a different reinforcement history of weather stimulus series. Much to his disappointment, the second experimenter finds that the original stimulus is either not effective or that another stimulus is more effective. Without realizing or accommodating the important role of conditioning, the controversy would be insoluble.

A concrete example of this problem might involve the following situation. Suppose a drop in temperature is a UCS for the elicitation of increased corticosteroid (hormonal) release from the adrenal gland. This response is not learned. Now assume that, due to weather dynamics, the drop in temperature is frequently preceded by geomagnetic disturbance. The geomagnetic disturbance could become a CS to elicit similar, although less intense, changes in corticosteroid levels. Without the close temporal proximity of the geomagnetic storm to the temperature drop, however, the geomagnetic stimulus might never be effective.

On the other hand, suppose in another part of a continent, along the ocean shore, for example, the drop in temperature is preceded by or is concurrent with an increased influx of some ion type. By repeated temporal pairing of the drop in temperature and the ion, the ion could become a CS. Although the ion was ineffective by itself, the effects of respondent conditioning render it a potent CS that can elicit a significant change in blood chemistry.

The existence of the conditioning variable within the biometeorological phenomena would be a constant source for failures in replication. An enthusiastic researcher could easily demonstrate persistent and reliable effects, usually of marginal statistical significance, between either of the two CSs, geomagnetic storminess or ions, and the blood measure. The researcher may even extrapolate to correlations between these measures and some secondary correlate of corticosteroid elevations, such as tendency toward diseases or stress effects. However, when this researcher went to the lab and exposed a group of subjects without a reinforcement history of either geomagnetic-temperature drop or ion-temperature drop conditions he would find no effect.

A competent behaviorist could discriminate many different stimulus patterns in the weather matrix that could be acquired by respondent conditioning. The person or population conditioned may not be directly aware of the conditioning experience. Instead it would rationalize or explain the effect from some cultural or personal bias. Here are some possibilities that have never been studied systematically.

The color gray is associated with depression or "blahness," with a concomitant complaint of lethargy. Such an emotional impression could be a learned consequence of a persistent stimulus pattern in the weather between low pressure air masses and cloudiness or grayness. If the drop in air pressure were an UCS that evoked a series of UCRs, such as slight depression, lethargy, or low-level inhibition, the color gray could become a CS due to respondent conditioning. Later if the color gray were presented by itself, it would be sufficient to elicit similar although less intense responses.

The color blue is paired with cheery, emotive conditions. People in North American culture frequently report increased ambulatory activity and general feelings of well-being when the sky is a deep, bright blue. Since this condition is frequently associated with high pressure air masses, especially during the fall and winter, and with cold temperatures, the color could become a CS. If the high pressure or colder temperature elicited responses associated with feelings of well-being and general excitement, the color blue, as a CS, would evoke similar low-level responses. The person may not feel euphoric when the color CS is presented, but he or she would display behaviors of similar polarity: cheerfulness or aesthetic preference.

Neither biometeorologists nor psychologists have systematically studied the role of respondent conditioning and passive acquisition due to the weekly presentation of the weather matrix. By the time a person reaches adulthood, when private experiences are presumably stable and emotional behaviors are well labeled, the person has been exposed to the complex changing weather matrix for hundreds of different episodes.

One can visualize this problem by picturing the human being or a population of human beings within a very large conditioning chamber: the environment. Even though people are mobile, they are still exposed more or less uniformly to the same stimuli since the weather stimuli occupy space dimensions much, much greater than they do. Within this chamber, the subjects are exposed to the complex weather stimuli. Each trial or presentation episode is comparable with the changing weather parameters.

The complex conditioning begins at birth, or perhaps even earlier, since the human fetus can be conditioned at least two months before birth. Since respondent conditioning involves predominately smooth muscle and endocrine responses, typically outside the awareness system, the person does not realize the actual pattern of the conditioning. Since the time factors for weather effects exist in from days to weeks, well outside the time experience of the present, the person would not easily find any connections. Various weather-related experiences could be explained in terms of what has been taught: the convenient catchall explanations of culture.

What are the Unconditioned Stimuli in Weather?

Certainly a first step in determining the role of respondent conditioning in biometeorological phenomena would be the isolation of UCSs. This essential problem has not been solved because by the time an organism, especially a human being, is old enough to be a subject, it has been exposed to a long history of complex weather patterns. Indeed, one could use snails or protozoa, but the generalization from these organisms to mammals is a tenuous and risky procedure.

Other difficulties contribute to the problem. Few data are available concerning which response systems are most likely to be affected as well as the latency or delay in reaction of these systems to a candidate stimulus. The data that are available suggest that a stimulus from the weather matrix would not influence all response systems to the same degree. Sound stimuli would be expected to influence different response systems than would electromagnetic stimuli. Changes in thermal stimuli would be expected to influence different measurable systems than would alterations in barometric pressure.

The optimal space of affect is complicated by a time variable. What is the typical latency of response in a system after the stimulus has been presented? In the laboratory situation, one usually presents a stimulus and records potential responses for a few minutes afterwards. Suppose the actual latency of a single response involves from hours to days following the presentation of the stimulus? Suppose the response is not a discrete event within an analysis interval of a few minutes, but can only be seen as an entity over temporal increments of several days?

Despite these problems, some guesstimates can be made about UCS candidates on the bases of statistical data and experimental work. One of the most persistent correlations in biometeorological studies is the triad: temperature, humidity, and barometric pressure. Statistically and analytically, it is very difficult to separate these three stimuli. In the laboratory, the relative effectiveness, in decreasing order of amplitude, appears to be temperature, humidity, and barometric pressure.

The close temporal proximity of these three variables is so persistent that they may at times act as synergistic UCSs, whereby the UCRs are even greater than the sum of the UCRs produced by each stimulus. With such potency, one would expect, indeed predict, that a wide variety of stimuli from the weather matrix could become CSs due to their close temporal association with the triad.

The UCS and CS problem is not an all-or-none solution. Conceivably, there exists a continuum of candidate UCSs within the weather matrix. Along this continuum lie most, if not all, of the major weather stimuli with varying degrees of intensity. In this situation, all the weather variables elicit their effects due to the genetic characteristics of the human being.

Alternatively, there may be only a few fundamental UCSs that correspond to the essential thermodynamic aspects of living systems. The continuum of intensity in this instance reflects only the degree to which the different weather stimuli are associated with the fundamental UCSs. Strong stimuli would be weather factors that are frequently associated with the presentation of the UCSs, while weaker or marginal weather factors would be those only occasionally associated with the UCSs. At present, there are insufficient data to determine which solution is correct or whether one solution is exclusive of the other. Both explanations could accomodate the available data.

A Radical Conditioning Option.

The necessary extrapolation of this dilemma would challenge the fundamental cornerstones of biometeorology. If the primary source of individual differences from weather effects lies in the different stimuli that have become CSs due to temporal association with a single major UCS, such as temperature, the repeated observation and implicit assumptions that a wide variety of weather stimuli produce similar effects would only reflect their association to the same UCS.

If this assumption were true, the hundreds of reports demonstrating an elevation of corticosteroids, for example, following exposure to electric fields, magnetic fields, ions, infrasound, winds, and so on would not indicate a general diffuse response to weather or a diffuse stress condition. Instead, the apparent similarity would indicate only that these stimuli had become at various times in different populations CSs to the same UCS that elicited elevated corticosteroids.

Due to respondent conditioning and respondent conditioning alone, these stimuli have become effective elicitors of conditioned responses. Without a history of temporal contiguity with temperature, these stimuli would never evoke the various organismic changes that have been observed. In this instance, the entire cluster of biometeorological changes, from the behavioral manifestations to blood chemistry, would ultimately be a consequence of a single environmental variable.

Other biometeorologists have suggested this possibility, following other assumptions or using alternative models. S. W. Tromp (1973) has suggested repeatedly in the last decade since the publication of his monumental text, that the single variable of temperature could account for the majority of the variability in biometeorological changes. Other variables in the weather matrix represent another factor or cluster of events quite distinct from temperature.

Superstitious Conditioning

Superstitious conditioning is an important confounding variable for explanations of the relationships between the person and the environment. Since the human being, scientist or not, tends to consider only those events of which he or she is aware, a thin and biased veneer of explanations is generated about causal relationships. The many events between the observed phenomena are not recognized, frequently not even acknowledged.

A simple example would be the sequence following the striking of a match. When asked, "Why did the match flame?" a normal person would say something like, "Because I struck the match on the package." In actual fact, depending upon on how detailed the analyses, the flaming of the match was due to a number of events that occurred in that split second between the strike and the flame. Since the person was not aware of these events, he or she could not respond to them.

The biometeorological situation is a quagmire of this type of conditioning. Suppose the person feels sick. There are a multitude of environmental events, of which he is not aware, that could have been responsible for this condition. The sick sensation could be due to the breakfast earlier that day, to some delayed response from a meteorological condition days before, or to some subtle behavioral stimuli that evokes diffuse feelings.

However, the explanation chosen would most likely involve some immediate obvious event in the environment. The person might look around and notice that the day is cloudy and conclude that cloudy days make him feel tired. Once that explanation is applied, it is likely to be used again. The label is reinforced not only by its coincident occurrence with the sickness, but also by virtue of the tendency for explanations to reduce the anxiety of uncertainty. Anxiety reduction can reinforce even the most farfetched explanation.

After this single, chance event, cloudy days will be perenially paired with that type of sickness. Since sickness is more effective as a reinforcer than its absence, the person is more likely to display the explanation "cloudy skies make me feel bad" primarily on days she feels sick. These periods, for most people, are sufficiently infrequent to prevent extinction of the explanation. Such extinction would quickly occur if the other reference, cloudy days, were used. The multitude of cloudy days in the year and the few sick days would quickly extinguish the connection.

The Time Factor

The optimal increment of time within which the body can respond to a stimulus has vague boundaries. Known events within the body have response durations of from a few milliseconds to several days. The latency of response onset to stimuli could be even greater, perhaps months or years. However, unless one assumes some type of memory analog in nonnervous systems, presumably the long latency would be mediated during the long period by a series of smaller stimuli.

In a similar manner, the mathematical limits of optimal stimulus durations are not clear. Laboratory durations of stimuli occupy time increments between a few milliseconds and a few hours. On the other hand, weather stimuli can be presented to the organism for from hours to days. Still other environmental stimuli may be available in temporal increments of from months to years.

Complexing this theoretical problem is the very nature of temporal contiguity, the bases of learning and conditioning. The close proximity of two events in time appears to be the fundamental requirement for their correlation within the body, specifically within the nervous system. The optimal quantitative dimension between the presentation of a CS and the presentation of a UCS had been assumed to be from about one to ten seconds.

However, these values have been reevaluated. Some stimuli, especially those associated with severe trauma to the organism and with stimulation of the autonomic nervous system, appear to violate this simple time interval. A single 20-minute consumption of a novel taste such as a sucrose solution, followed six hours later by a singular sickening episode induced by the novel or toxic substance produces a conditioned taste aversion to the substance.

The critical theoretical question: what has mediated the sensation of the new taste long after it has been gone, since the animal did not know it would be sick six hours later? Without alluding to mystical concepts, one must assume that some reminiscent characteristic of the new taste existed for sufficient periods for the pairing to occur.

Available data indicate the autonomic nervous system as the likely mediator of the stimulus. At the chemical level, the process appears to be related to some control over the degree of protein synthesis in the brain portions (limbic system) associated with emotional memory. More detailed questions concerning the type of protein synthesis influenced or the severity of alteration cannot be answered at this time.

In principle, the involvement of these two systems significantly increases the increment of time by which two separated stimuli can be correlated. Experiments with rats indicate periods of six hours or more for smell-taste stimuli. However, drug experiments involving the interference of protein synthesis after single learning episodes indicate that the labile process associated with conditioning is present for about three to four days.

Assuming these rodent data are applicable to humans and the weather matrix, the conditioning problem may be more important than previously suspected. Depending upon the nature of the UCS, a component of the weather matrix could become an effective CS even though it was presented and terminated days before the actual presentation of the UCS. The role of the autonomic nervous system and its interface with the locus ceruleus (the dream center) and other brain stem nuclei associated with protein synthesis would acquire special relevance.

LANGUAGE AND THINKING AND THE BODY LOCUS

Words and sequences of words are powerful stimuli that can evoke responses throughout the body and modify behaviors that pre-

cede them. Four-letter words can cause fainting in some people.
A simple statement, such as, "Your mother just died" can evoke
changes not unlike a full-blown anxiety attack. To produce similar
changes with nonverbal stimuli would involve environmental energy
magnitudes much greater than those required to state or read the
sentence.

Language as Measurements of Events

Although words may be powerful, they are limited measurements of environmental events. Language is in the least a nominal
scale and an ordinal scale at most. Nominal scales are all-or-none
types of labels, such as black versus nonblack or cloudy versus
noncloudy. Usually, nominal words appear in dichotomies, such
as black versus white, right versus wrong, yes versus no. There
are no in-betweens with this type of measurement.

The limit of this scale should be evident since there are only
two options: the stated one and its opposite. The opposite may not
be stated, but is implied. For example, the statement, "I feel bad
when it rains," implies that the second condition is "not rains."
In other words, there are only two meteorological conditions—rain
versus nonrain—as the environmental label.

Nominal scales encourage the inclusion of heterogeneous
events within gross dichotomous labels. Essentially, this type of
measurement divides all possible environmental events into two
categories. If there are only two weather descriptions—bad and
good—hundreds of events are separated by these two all-inclusive
labels. By this process, the speaker often assumes that all events
within the good category originate from a single source, while all
events within the bad category originate from a different single
source.

By using nominal word scales, one is restricted to a primitive
differentiation of the many complex parts of the environment. Because of the limited measurement, all types of different weather
conditions—perhaps thousands, depending upon the sophistication
and resolution of the measurement scales—will be called good or
bad. There is no option to distinguish the different types of weather
within the category.

The problem is similar to medicine in the late nineteenth
century. During this era, all lung ailments—tuberculosis, cancer,
the common cold, influenza, and pneumonia—were collectively called
vapors. With this diffuse, overly inclusive category, differentiation
of the different types of lung ailments was difficult. Furthermore,
the application of a single drug, such as an early antibiotic, primarily

influenced the recovery of pneumonia patients. Since other patients diagnosed with vapors would not respond to the drug, the treatment was rejected or considered ineffective.

More sophisticated language measurements are ordinal scales. Each increment is a whole concept along a particular dimension. An example of an ordinal scale is ranking characteristics, either with a number or a word, from least to most. An example would be something like good, better, best. An alternative, involving numbers, would be to rank ten people from best- to worst-looking from one to ten.

Using ordinal scales, one could also rank subjective states, such as pain or anxiety, along a zero (no pain or anxiety) to five (maximum) scale. The major limit of this scale is the unequal nature of the intervals. There is no way by which one can determine, by using the scale, whether the intervals are equal. To state that a pain rating of three is three times more intense than a rating of one is not valid.

Language as a Label of Organismic Events

Within its measurement limits, language can be used as a label for environmental events. The components of language, words and syllables, are discrete events that concisely exist within perceived time. Objects or events that are clearly defined in space or time are best described by language. Such events are easy to verify, count, and define.

The immense list of discrete object-events comprise, for the most part, the indisputable aspect of human culture. They are easily labeled and can be identified, counted, and agreed upon indefinitely. Such stimuli include chairs, tables, people, cars, books, and such objects and walking, talking, sex, and such events. The common property of these object-events is some independent reference outside the body locus. They are perceived within specific time intervals.

Discrete words or labels for these object-events are biased within the visual and auditory modes of experience. Object-events that can be seen or heard comprise the major stimuli about which people agree. Concerning a given object, such as this book, most people with normal vision can not only agree about what the object is but can also describe some of its fundamental characteristics.

Words as reliable measures of human experience begin to lose their potency when they are used to describe events occurring within the body. Events or objects associated with touch, proprioception, and temperature are more difficult to specifically label with words. Stimuli influencing taste, smell, and internal senses such as autonomic tone are even more difficult to reliably describe with words.

The problem lies in two parts. First, words are discrete events optimally used to describe discrete occurrences. Diffuse sensations or conditions within or outside of the body cannot be easily accommodated by a labeling procedure. Word labeling is limited by the ability to discriminate a specific space and a particular time increment.

Conditions that are everywhere and amorphous are difficult to break into smaller parts or discrete elements. Words for these conditions, such as field, spirit, emotion, god, and sometimes even weather, create great difficulties of resolution. Conditions that are everpresent and amorphous are also difficult to break into discrete elements. If the beginning and end of the event fall outside the human time reference of now, the event becomes confusing and anxiety-provoking. Words for such conditions include infinity, eternity, and, to some extent, time.

The second problem can be traced to the general paradigm of learning language. In the general development of language skills, the child first learns nouns (for objects), then verbs (for action sequences or events), and then words for different human references (pronouns). The process is essentially a pairing procedure by which the parent or teacher shows an object and immediately pairs it with the appropriate word.

The parent picks up a chair and says the word "chair." As a result, the object and the word are paired together in space-time. If several children are involved, all the children can learn the word "chair" as a description or label for the object. To maintain this behavior, initially at least, the parent usually follows the correct or appropriate statement of the child by some type of reward, such as a smile or the phrase, "That's good."

Through this basic procedure, the child-adolescent-adult learns to pair objects and events with words. As the person grows and acquires more and more complex associations, pairings can be laid upon pairings. By adulthood, the average person can display a large repertoire of labels, such as, "This behavior is schizophrenic," or, "This is a drive shaft from a 1976 Ford," or "This is the dorsal region of the ventral medial hypothalamus of a 30-day-old rat brain." In the vast majority of instances, people with similar labeling histories would agree.

At about the same time language usage proliferates in the young child, private behaviors develop. The intimate relationship between verbal or vocal conditioning and thinking is well known, but is beyond the scope of this text. In general, the principles associated with the acquisition of language are very similar to those associated with the characteristics of private behaviors.

After about four years of age, the child begins to report things happening in his private behaviors. The child's verbal behavior indicates the development of an adultlike memory and the beginning of phenomena that we infer to be images, thoughts, and ideas. It is also apparent that the child may not always discriminate between images induced by objective stimuli, such as watching a thunderstorm, and dream images of a similar theme. Both may appear real.

How does the child learn to label his private behaviors? How does she learn to describe her emotions? In principle, the same procedure is used for private behavior labeling as for public behavior labeling: the pairing of word with event. This operation is done so routinely by parents and teachers that it is probably not recognized as singularly important.

When the young child approaches the parent or teacher and begins to stutter in some context, the adult offers a label to the child. After the child slams his finger with a hammer and begins to cry, the mother says, "You are hurt." When the child wakes up in a startle at night and says a large monster was creeping on the floor, the father says, "No, you were only dreaming." Slowly, word labels for private behaviors are paired.

Pairing of words or phrases to private experiences is neither systematic nor reliable; it can't be. Most labels for a child's private behavior are learned from the parents' or teachers' inferences about how the child should be feeling. Since there are no independent references for the private behaviors when labeling is taught, the teacher must infer how the child is feeling or what the person is thinking about by external cues.

Imagine the following situation. Young Johnny is looking out the window. He is six-years-old and has been going to school for five months. Instead of his usually vigorous after-school episode and voracious appetite, he stares out the window and doesn't eat. To Johnny, the sensations are odd and peculiar; he has never experienced them.

The concerned mother watches the young boy for a few minutes and then asks definitive questions. "Do you have a fever?" "No." "Did you lose your friends?" "No." If, by chance, she then says, "Well, you must be in love," she has just labeled young Johnny's private experience. From this point on, whenever this or similar private behaviors recur, Johnny will likely say or think that he is in love.

Clearly one can see the built-in error within this unavoidable type of learning. Although the same label of love is used by millions of people, the actual series of private responses to which it refers may be as variable as the number of people involved. Whereas Johnny may have actually been ill during the time of the initial pairing of the private condition with the word love, other children may have been displaying a multitude of other conditions.

Types of Descriptors

The individual meanings of words used to describe or to remember emotional behaviors are as diffuse as the events they describe. Metaphors, similes, and references to external processes are the most common verbal descriptors of emotional experiences. People report that they feel "like a limp wash rag" when they are tired and asthenic. Some people report "they are tight as a rope" when they are tense. Responses with a clear or periodic time component are often coined in terms of dynamic events, such as "feeling like a butterfly in flight" or "a cat on a hot tin roof."

People learn to use words such as bad, great, good, and fantastic to describe a series of complex, diffuse experiences. Implicitly, since a single word is used, people begin to assume that the behaviors involved are from a singular, discrete source. One assumes that the processes resulting in the label "bad" for one day are the same as the processes resulting in the label "bad" for another day. Combined with the normal egocentric reference for experiences, a single-label language for emotional changes can be very misleading.

In many experiments, researchers go beyond the response limitation and conclude that if thousands of people report feeling bad, lethargic, odd, or some other label for diffuse processes, all these people are from the same physiological population. In other words, researchers may assume implicitly that all statements of "bad" are descriptive of the same events and the same mechanisms. They mistake reliability for validity.

When this occurs, the problems of the vapors error are resurrected. Since all people who say they feel bad on a given day only appear to be similar, because of the vague label, but in actual fact are reporting this behavior because of many different stimuli, single analysis or treatment will never be possible. As in the vapors problem, a wide variety of causes will be indiscriminately grouped within the same label: "These people feel bad."

As a result, correlation of days of feeling bad with some single weather variable, lagged or not in time, must always produce results of marginal, if any, statistical significance. The majority of cases of a single analysis will involve many different mechanisms responsible for the feeling. They will have different sources of variance that reduce the relative intensity of any given effect. Biometeorological data based upon verbal responses and descriptions of private behaviors will always be the heterogeneous pool of great variability.

If there are many different mechanisms functionally associated with the same verbal label, such as bad feelings, weak correlations exist persistently in data based upon human self-evaluation. If there are many different physiochemical patterns that are made similar only by the superficial similarity of the word used to describe them, one would expect, statistically, every major weather variable to influence at least one of these mechanisms. Since each of these mechanisms will be described by the same vague label, low-level correlations will persist. They will be ubiquitous but never tractable.

Problems of Discrimination

The failure for words to aptly describe private behaviors, such as emotional experiences, applies not only between, but also within, individuals. Most untrained people cannot easily differentiate subtle changes in the autonomic nervous system. Only intense emotional behaviors with clear contexts are distinguished.

Although there may be hundreds of different autonomic patterns to as many environmental stimuli, the number of words developed in this culture to describe autonomic experiences are limited to only a few. By necessity, many different stimuli or effects from different stimuli will be subsumed under the same experience or word. Even though these stimuli have nothing in common in terms of source, their similarity is derived from the system of influence (the autonomic nervous system) and the word descriptor.

The consequence should be evident. An average person just does not have the discriminatory apparatus to tell which source is which when vague experiences are involved. Stimulation from weather changes, subtle social stimuli, fatigue, low-level infection, emotional upset, and perhaps even aging processes will all elicit similar experiences.

Any researcher who must use human verbal behavior as the primary measure of autonomic behaviors will be faced with this problem incessantly. He or she will be plagued by overinclusion of stimuli that are not originating within the weather matrix. Since the assumed weather effects and the consequence of the autonomic nervous system are so diffuse, even the measurer (the personal observer) may not be able to discriminate the source of these changes.

Words as Determiners of Behavior

Words are not only descriptors of behavior; they can also be determinants of behavior. The public (vocal) or private (thought) symbol or word has a high probability of display whenever the object is presented. If the word "book" is paired with the object book, the person soon begins to emit the word "book" or to think about books (a private verbal response) when books are presented. Words can become stimuli that evoke public and private responses.

Behaviors can be labeled in a similar manner. A particular sequence of responses may be called stupic, while still another series of responses may be labeled intelligent. The whole human organism can be labeled as well, from the specific name—Jim or Jill—to some other social descriptor—idiot or genius—or to a deprivation condition—piece or hunk.

These general labels can determine specific behaviors if they are applied to the person frequently enough by a reward source (parent, teacher, or even peer group). If a person is called stupid frequently enough, he will soon begin to display the behaviors associated with a stupid person. He may even begin to think of himself as stupid. Usually this conditioning is expressed when the person begins to label his own behavior as stupid. The conditioning is complete.

In a more complimentary nature, positive labels and positive behaviors toward a person can change behavior in the opposite direction. Some years ago, a famous psychologist selected children with average intelligence quotients (IQs) (100) from grade schools within a city. These experimental children were then transferred to different schools. Before they were introduced to the new class, the psychologist told each teacher and class that the new student was a gifted genius and that the purpose of the study was to determine how young geniuses behave in a normal situation.

During the school year, the experimental students were treated as geniuses by both teachers and students. The odd verbal behaviors, peculiarities, and comments by the experimental students were given special attention by the teachers and students. In fact, the other students began to expect the experimental students to show accelerated or superior learning.

When tested again at the end of the year, the experimental students demonstrated an average 20-point increase in the IQ scores, an extraordinary increase. In addition, their behaviors had changed markedly in the direction of how they were rewarded by the class and teachers and how they anticipated a genius should behave.

The total impact of a verbal label upon a person's behavior is fundamentally related to a person's reinforcement history. Since most of the words for a person's private responses are derived from a parent figure, it should not be surprising that people continue to depend upon other people for descriptions of their behaviors. To some extent, all of our private behaviors are a function of other people's verbal behavior.

The large range of behaviors that can be evoked by the appropriate verbal behavior is amazing. Some people will experience deathlike conditions after hearing a lecture on death. Medical students sometimes break out in the symptoms of typhoid after hearing a lecture on the disease. After reading about tetanus (lockjaw), one woman developed its symptoms following a minor cut; she actually displayed seizures even though there was no organic evidence of the toxin.

THE PLACEBO OR EXPECTANCY EFFECT

The power of words becomes clearly evident with a phenomenon known as the placebo or sugar pill effect. By definition, a placebo is a pharmacologically inert substance that evokes clear behavioral or physiological changes. Since the placebo presumably has no known physiological effect, its major impact appears to be related to the instructions and psychological expectancy associated with the placebo treatment.

Expectancy and Private Behaviors

The placebo effect is a special condition of the more general phenomenon of expectancy effects. Expectancy effects are behavioral consequences that are produced by labeling a situation, usually one predominated by ambiguity or anxiety, with a particular word. The more emotionally loaded and anticipatory the word, the more effective is the expectancy effect.

An example of the expectancy effect follows. Suppose there is a constant background of thumps in your house, due to temperature change effects. During the day you dismiss them as trivial. One night you are alone, reading a murder mystery. The words are ordered in such a manner that autonomic arousal has become more and more intense. Suddenly, one of the normal thumps startles you.

If sufficiently aroused, you would normally include the mundane thump within the experiential framework—the structure of the verbal situation. The thump would no longer be just a thump, but would become a footstep; indeed, the more you thought about it, the more it would seem to be a footstep. Some people would feel uneasy the rest of the evening or quit reading until someone else came home, while others would continue to greater intensities.

In the latter situation, the plot of the book could determine or influence more and more of the private behavior. Tingling sensations would occur at the back of the neck and top of the head. The

person might report goosebumps or the feeling that someone or something were in the house. In more extreme instances, the person
may actually see a human face—markedly similar to that imagined
or described by the author—for a brief moment.

These changes are a consequence of expectancy effects. They
are neither trivial nor limited to just thinking. For many people,
word sequences are sufficient to induce sweating, heart rate changes,
and blood chemical alterations similar to those produced by more
physical stimuli. In some cases, expectancy do-loops (repeated
thinking of themes whose effects increase with each sequence), can be
deadly.

This principle has been applied to emotional behaviors in
general by Schacter and Singer and their colleagues (1962). Their
data indicate that the cognitive aspects of the situation determine a
significant portion of the emotional experience. Even though the
physiology may be similar, the actual details and specific emotions
reflect the social structure or expectancy (the cognitive aspects) of
the situation.

Briefly, to demonstrate this effect, these experimenters injected
human subjects with adrenalin. Each subject was placed in a room
where a stooge (a person involved with the experiment) acted either
angry, happy, or complacent. After a few minutes, the experimental
subjects ranked or reported their experiences. Even though they
were injected with the same substance, their private behaviors reflected the social-verbal context in which they had been placed.

Subjects that had been placed in rooms with angry stooges
reported anger, while those placed in happy situations reported joy
or euphoria. Not only were the verbal behaviors determined by the
context, the emotional behaviors appeared to be influenced as well.
Some angry students "wanted to fight," while others "felt perturbed
and uncomfortable for the rest of the day." Some happy students
were giddy, to the point of a "high."

Expectancy of relief is a primary factor in the effectiveness
of many so-called therapies, such as meditation. The anxious or
expectant subject anticipates, similarly to the person reading the
murder mystery, the types of private behaviors associated with the
therapy. Concepts that embody diffuse phenomena are very effective.
Diffuse occult and mind-expanding themes breed a complex of fantasy
and anticipation; they are also prone to superstitious conditioning.

Expectancy effects are well known to influence the private
experiences of weak psychotrophic drugs such as marijuana. Many
experiments have indicated that experienced smokers merely exposed
to the smell of similar smoke within the usual social setting report
a similar high even though no active ingredient is within the "weed."
Whereas experienced smokers may give the substance a "high" rating,

novices, those people who have no expectancies associated with marijuana, report little effect.

Egocentricity and Explanation

The average person assumes that he or she is not only unique but is an exception to the rule. Despite statistics that indicate increased higher mortality or susceptibility to disease, many people continue smoking cigarettes or drinking alcohol in excessive amounts. The problems always happen to the other guy.

This normal behavioral characteristic becomes especially evident in situations in which private behaviors have been modified by expectancy effects. Since the person uses his or her private experiences as a reference point, indeed many times as a proof of environmental events, he or she insists on the validity of private responses.

The theme is old and repetitive. Private experiences of each individual always seem a little more real or better than another person's. Other people can be fooled by the situation, other people can be gullible enough to misinterpret experiences; other people can be naive enough to experience a high when there is no chemical source. How many times have you heard a statement such as, "Weather effects only sickies and hypochondriacs—it's all in their heads—but in my case, you know, I'm really sensitive to . . ."

Placebos and Personality

A normal characteristic of early childhood is the capacity for adventurous and imaginative involvement. Children normally display periods of imagination during which they talk to invisible playmates or become so involved with the context of the situation that they report a feeling of being in the context. They often become emotional in these situations, even to the point of crying or intense autonomic changes.

Typically, this tendency decreases with age. When it remains functional into adulthood, hypnotic manipulations appear more frequently and with greater ease. Involvements that maintain this ability in people are reading, drama, creativity, religion, and a sense of noncompleted adventurousness. Mysteries, adventure stories, biographies, and science fiction allow a rich fantasy of participation in word sequences. In other words, the word sequences in the book begin to control the direction and emotion of the person's private behaviors.

During the involvement with reading or worrying about some context, hypnotic-related responses are typical. People prone to this behavior report that the environment fades away and the story characters or the context become real. Intense feeling of heat, cold, pain, and sexual arousal can be elicited by the sequence of words (the context); some may actually have hypnotic dreams.

Accumulating evidence has indicated a close similarity between people prone to placebo effects, expectancy effects, and suggestibility—the probability of being hypnotized. Although the mechanisms are not clearly isolated, one evident fact is the superficial nature of the trance state. Apparently, a suggestible person does not have to be in a trance in order to display suggestible-related behaviors. The trance condition—rigid body, staring ahead into deep space, and so on—appears to be an expectancy of how to behave when hypnotized, acquired through movies, books, and other media.

The existence of suggestible people is an expected human behavior considering the total dependence of a child's labels for internal experiences upon the parent. It is predictable that some children continue to depend upon others, especially parent figures, to control their private responses. These individuals, who maintain childlike psychological characteristics, are directed by parent surrogates, such as doctors and religious leaders. Their experiences may also be determined by what they read.

People mildly prone to suggestible behaviors, roughly 40 percent of the population, display a number of personality patterns. The most general theme that permeates their symbolic (thinking) responses is a deep core of basic beliefs with which great emotion is associated. Predictably, these people are likely to be more religious and to be members of sects prone to spontaneous conversions. Similar to placebo reactors, they are prone to be anxious, dependent, self-centered, and preoccupied with internal bodily processes. Frequently, they are seen by others as quiet, brooding, and resentful compared with nonreactors.

People greatly prone to suggestible stimuli or instructions also show a readiness to trust, an acceptance of logical incongruities (a rigid scientist might be intensely religious), a relative suspension of critical judgment (even though the relationship between a belief and a behavior may be demonstrated as a chance occurrence, they still assume there must be something there), an overall obstinacy an excellent memory, a rigid core of private beliefs, and a subtle sense of inferiority. They comprise not more than 10 percent of the population and were once called hysterical patients.

Clinically, these symptoms can be repressed or denied by a thin facade of scientific jargon or general logical thinking. However, these rituals of responses are easily dissociated or disrupted when the person is challenged on a personal basis. Neither intelligence nor education protect against the personality characteristics of suggestibility. A suggestible Ph.D. is little different on these dimensions from a suggestible person with a first-grade education.

Diseases Treated by Expectancy

There are discrete clusters of disorders or symptoms that are prone to alleviation or removal by placebo procedures. They include some types of headaches, neuritis, tension, some arthritic complaints, menstrual cramps, lethargy, inability to concentrate and, most of all, idiosyncratic pain. All of these symptoms can be alleviated, some to total improvement, with appropriate instructions or expectancies.

These symptoms as well as the syndrome of neurosis are loaded heavily by anxiety or some negative anticipatory component. Many times the anxiety may be produced by a vague feeling of insecurity or a general experience of inability to structure or predict the environment. If the major source of the failure is associated with predicting the future and the futre is associated with negative possibilities, such as the end of the world or personal death, the anxiety can be permanent.

In these contexts, any series of words or rituals that structure the environment can produce a reduction in the anxiety and the symptoms associated with it. The most effective expectancy-placebo treatments usually give the person a general plan for the future as well as a structured pattern for everyday living. The long-term plan may include anything from knowing how one will die to the expectation of rewards. The daily ritual may vary from pill taking with no pharmacological consequence to prayer to periods of silence. The structure is the important factor.

In light of the marked individual variation in the types of stimuli that contribute to anxiety, one would expect a rich variety of different treatments optimally suited for the different events in personal histories. Some people feel better just by visiting a physician, especially if there is a great deal of ritual and impressive machinery, while others improve after trying some popular therapy.

Physiological consequences of placebo treatments are as impressive as the behavioral reports. Suggestible people can display wheals and changes in skin color following appropriate suggestions. The onset of the color changes may occur hours later. Manipulations of blood pressure, consciousness, and the perception of pain are frequent responses. The more impressive the situation and the therapist or hypnotist, the greater, in general, the effect.

In one well known study, a group of medical students were told that blue pills were tranquilizers and red pills were stimulants. They were instructed to take the pills as they wished, although they were experimental drugs. In actual fact, both types of pills contained inert substances.

After a few weeks of consumption, students reported that the red pills made them active and alert while the blue pills induced relaxation. In addition to these reports, over 55 percent of the students demonstrated changes in blood pressure in the direction of the pill that had been taken.

The final, most powerful, single category of behavioral events influenced by expectancy is the subjective evaluation of pain. Subjects who are told that an unknown type of shock may be unpleasant rate the subjective amplitude significantly higher than subjects receiving the same actual current but who have been told to expect a simple sensation. Presumably, the increased subjective experience is a function of anxiety: the anticipation of something bad.

Application to Biometeorology

There is an unfortunate propensity to separate psychology from biometeorological phenomena. Usually, psychology is reserved for cases that are purely subjective or superficial. Implicitly, many biometeorologists and physical scientists assume that a real phenomenon does not contain a significant psychological component.

As a result, few studies have considered the contribution of conditioning over a lifetime or of public or private verbal behavior to the display of biometeorological responses. This becomes important when one realizes that physiological changes and alterations in blood chemistry can occur with psychological stimuli alone.

Complex human behaviors, such as anxiety, that can be generated by words or symbols have been ignored in many biometeorological phenomena. Behaviors determined in large part by expectancy have been considered in clinical situations and in subjective experiences. However, the role of this factor in discrete data such as accidents, has not been mentioned. A frequent presumption is that clinical complaints and weather are different from discrete events, such as traffic accidents and weather.

Yet the languages of most human cultures contain clichés and platitudes about weather and behavior. These phrases are repeated almost reflexively to explain and describe behavior. Word sequences such as "Gray days bring you down," "It was a cloudy day when he died," "May breeds suicides," or "People go weird on full moons" are peppered throughout everyday language. In light of the general tendency for suggestibility in human subjects, how many accidents are a consequence of the expectancy associated with weather types?

Critical questions must be asked about biometeorologists' favorite test animal: the human being. Are people who are weather-sensitive, also placebo reactors? Do weather-sensitive people have different scores on personality tests? How does one separate psychological stimuli from physical stimuli? Are they really separable within the biometeorological context? With these reservations, one can proceed to the detailed responses associated with weather.

5

TYPICAL WEATHER CORRELATES AND PROBLEMS OF CLEAR DEMONSTRATION

Human responses most commonly correlated with weather matrixes share a general cluster of properties: they are diffuse, difficult to quantify, apparently inconsistent, and produced by a wide variety of other everyday factors, especially psychological stimuli. With the exceptions of accident and hospital admission data, most weather measurements are dependent upon the arbitrary, unstable, and limited instrument of human experience.

Since people evaluate weather effects according to a very limited measurement scale, such as "I feel better" or "I feel worse," detailed analyses and, hence, greater understanding of weather-biological mechanisms are impossible at present. Even discrete data, such as the number of accidents during barometric pressure changes, are subject to crude analysis. Frequently, experimenters divide weather periods into simplistic halves—days of increasing barometric pressure and days of decreasing barometric pressure—without reference to specific numerical changes.

As a result, most weather correlations with human behaviors are vague and typified by statements such as "More suicides occur during warm fronts" or "More migraines occur during damp cold weather." To state that something is more or less likely to occur smacks of flipping a coin. One must know how much a particular behavior increases or decreases. Do these responses change according to a quantitative alteration in the weather matrix? Are there other variables, not measured or reported in the study, that could account for the alleged single-variable effect?

GENERAL PATTERNS OF HUMAN WEATHER
RESPONSES

Not surprisingly, the general pattern of weather complaints
and responses is coupled with changes symptomatic of autonomic
activity. Very similar changes are associated with lunar phases,
geomagnetic storms, and other exotic stimuli. Whether this pattern
reflects a real diffuse mechanism capable of responding to the vast
range of subtle environmental stimuli or, instead, merely empha-
sizes the crudeness of our measures remains to be established.
Since diffuse responses are difficult to manipulate experimentally,
they often remain reservoirs of emotional proofs for beliefs, super-
stitions, and pseudoscience.

Typical Behavioral Complaints

Verbal behavior, the response sequences a person says or
writes, is the primary source of weather data. Symptoms said to
be intensified by the weather matrix are tiredness, discontent (with
the world and self), dislike for work, inability to sleep, restless or
interrupted sleep, failure to concentrate (especially difficult discrimination, such as learning new material), headaches, general nervous-
ness (fidgety or apprehensive responses), depression, forgetfulness,
aching of healed fractures or surgical scars, vertigo (or fainting),
visual flickering (black spots in front of the eyes), disturbances in
circulation (especially the extremities), chest pains, and, or course,
rheumatism (joint pains).

These complaints have been reported by different-aged populations of presumably normal people both through metaphor and by
questionnaire sampling. The syndrome (specific collection of symptoms and complaints) may vary in direction and intensity with the
type of weather change. Cold fronts, warm fronts, and special
winds may elicit different patterns. However, some variant of these
complaints appears to comprise the major profile of response.

Less frequent complaints allegedly linked with weather changes
are moist palms (or sweaty palms and feet syndrome), increased
perspiration, shivering fits, loss of appetite, diarrhea or constipation, alterations in sensory input (especially hearing, taste, and
smell), and hypersensitivity of the skin. It is not clear whether they
are elicited by weather changes or whether certain people, prone to
weather sensitivity, already have these disturbances. The weather
would merely unmask these propensities.

Typical Discrete Behavioral Changes

The second category of weather-related behaviors involves
people who display some response rather than report private experiences or feelings. These measures are all-or-none; they either
occur or do not occur. They are usually reported in percents of
total numbers of occurrence above or below some expected frequency,
such as chance. Typical measures have been numbers of job or
school absentees, traffic accidents or deaths, industrial deaths or
accidents, suicides, homicides, and births.

It is unlikely that weather stimuli only influence the numbers
of crises or deaths in human populations. The tendency for weather
effects to be paired with these extreme events is more likely a consequence of their psychological and economic impact upon human
experimenters rather than a special property of weather. No doubt,
there are many, many other trivial behaviors whose numbers or
rates of occurrence are correlated with weather. They would not
be documented as obsessively as are deaths and crises.

Biological Responses

A wide variety of biological responses and medical complications have been blamed upon or correlated with weather changes.
These effects include alterations in the occurrence of migraine
headaches, strokes, epileptic seizures, glaucoma pains, joint swelling (objectively measured), urination, blood clotting, blood volume,
skin conductivity, tissue permeability, embolism risks, blood clots
in lungs and legs, thyroid activity, heart rate, blood pressure,
heart attacks, and toxicity to certain stimulants.

Relationship to Specific Weather Changes

The biometeorological literature abounds with contradictory
patterns. Scarcely any general statement can be made concerning
weather change correlates without at least one empirical exception
to the rule. At present, we cannot conclude objectively whether
these contradictions reflect the nature of the phenomena or the
indiscriminate lumping of different weather conditions and different
human populations.

Some studies report human behavioral-biological changes
following any change in weather, while others differentiate only between cold and warm fronts. Still other studies select only one
member of the famous weather triad of humidity, barometric pres-

sure, and temperature and ignore the other two. These studies conclude that organismic responses occur when the barometric pressure is falling, for example, without reporting temperature and humidity changes at the time.

Despite these limitations, a few general patterns emerge. Considering the great variation within the biometeorological literature, these patterns may not be trustworthy. In general, decreased temperature, increased barometric pressure, and increased humidity are associated with sympathetic stimulation. Consequent or correlated symptoms are increased urination, more severe kidney pains, migraine headaches, peripheral circulatory complaints, joint stiffness (as opposed to pain associated with water retention in joints), and general increased muscular activity strength. Psychological complaints during warm fronts with falling barometric pressures, from pain to depression, appear to be reported less frequently or to be less aversive.

Influxes of warm air accompanied by falling barometric pressure and high humidity, as fronts or as thunderstorms, have been blamed for a much greater cluster of biological adversities including water retention, restlessness among psychiatric patients, arthritic pain associated with swelling, depression, irritability, and rashes of various accident forms, from traffic to industrial. Disorders antagonized by water retention, such as glaucoma, menstrual swelling and varicose veins, are said to increase during these periods as are hemorrhaging complications. Most of these effects are symptomatic of parasympathetic activity.

Although a simplistic approach would encourage the assumption that cold fronts/air masses stimulate the sympathetic nervous system primarily while warm fronts/air masses stimulate the parasympathetic nervous system, the problem is more complicated. This speculation is evident only at a superficial level. Any weather change that stimulates an unstable autonomic nervous system would evoke overshoots and undershoots involving both parasympathetic and sympathetic complements. Autonomic tone (the balance between the ortho- and parasympathetic responses), baseline at the time of the weather change, and learning histories of individuals or populations must be included.

Magnitude of Weather Effects

In absolute terms, the effects of weather change upon the human population are very insignificant. This fact must be remembered otherwise misleading exaggerations can arise. Except for the blatant turbulent forms, such as tornados, most weather-related events evoke changes that are barely detectable above the trials and tribulations of daily life. Domestic quarrels, a vacation, a good party, and, to some people, even an emotional movie or soap opera, evoke greater behavioral-biological changes than does the weather.

Five recent medical and psychological studies clearly demonstrate that weather variation contributed to no more than 10 percent of the total complex of stimuli influencing human mood and behavior. Since typical daily variations in average human behavior, from moods to medical complaints, display fluctuations between 10 percent and 30 percent above or below the monthly baseline, weather effects are well-immersed in the background noise of living. These data do not deny weather changes as important stimuli. Rather, they place them in an appropriate perspective.

GENERAL METHODS OF DEMONSTRATION

The methods, logic, and measurement tools used to isolate and interpret a phenomenon determine the nature of the phenomenon. Any effect can be overlooked or exaggerated, depending upon the techniques of detection. Since methodology is so important, especially in complex areas such as biometeorology, close attention should be paid to these procedures. A few of the problems and limitations of classic weather-organism relations are described.

The Correlational Method

A correlation between two variables, such as ambient temperature and reports of human mood, indicates that change in one is associated with a change in the other. If both variables change in the same direction, the correlation is positive. If one variable increases (for example, a rise in temperature) while the other variable decreases (for example, a lowering of mood ratings), the correlation is considered negative.

Statements such as increase or decrease are only qualitative measures. They do not tell how much change occurs or to what extent the two variables are related. Such data are required for everyday predictions. To determine the quantitative relationship between two variables, numbers must be used. One must compare the numerical range in one variable, such as from 10°C to 30°C, with the numerical range in the other, such as mood ratings between 1 and 10.

The quantitative relationship between two variables is typically given a numerical value between 0 and 1 or 0 and -1 (for negative correlations). The value is called a correlation coefficient. As the coefficient approaches 1 or -1, the relationship between the two variables becomes greater and greater. One can predict more precisely the value of one variable from the value of the other.

Different types of correlation coefficients are used for different types of measures. There are four basic types of measures or scales: nominal, ordinal, interval, and ratio. Nominal scales divide events into all-or-none categories. For example, something is either green or it is not green; something is either good or not good (implicitly bad). Any dichotomy based upon opposites, such as good versus evil, God versus devil, or right versus wrong, in which there are no intermediate options is an example of a nominal scale. If a person either feels good or bad, statements about mood would be nominal measurements.

Ordinal scales rank events along some dimension. If you rank five people according to weight, from heaviest to lightest (either 1, 2, 3, 4, or 5), an ordinal scale is involved. With ordinal scales, the intervals between ranks are not equal. A person who is ranked second is not one half the weight of the person ranked first. If the five people weighed 200 lbs., 190 lbs., 103 lbs., 102 lbs., and 100 lbs., the difference between the people ranked second and third is not comparable to the differences between the people ranked heaviest and second heaviest or those ranked lightest and second lightest. Rankings are not absolute measures.

Interval scales assume equal intervals between units of measure. On an intelligence quotient (IQ) test, the difference between scores of 100 and 120 is 20 units. These 20 units are considered equal to the 20-unit difference between scores of 120 and 140. However, the scale does not allow the conclusion that a person with an IQ of 200 is twice as smart as a person with an IQ of 100. Ratings of private experience, such as mood or the intensity of a stimulus (light, sound), can be interval measures if many different measures of the experience are made with the same numerical scale. If all moods are ranked from lowest to highest, the scale must be considered ordinal, not interval. If daily moods are given a number (between 1 and 10, for example), an interval scale is involved.

Ratio scales are more common in the physical sciences than in the behavior sciences. Ratio scales have equal intervals between units and an absolute zero. Temperature (° Kelvin) and length (meters, leagues, parsecs) units are examples of ratio scales. One can legitimately say that two meters is twice as long as one meter, but one cannot conclude that 40°C is twice as hot as 20°C. The latter scale does not have an absolute zero; the scale has negative values.

There are great differences in the predictive capacities of nominal and ordinal scales and interval and ratio scales. For all

practical purposes, the first two are qualitative while the second
two are quantitative. Verbal behavior and all psychological experiences measured by words are fundamentally nominal-ordinal scales.
Human beings perceive and label most events in the environment by
these scales.

The Pearson Product-Moment Correlation

The Pearson product-moment correlation, designated by the
symbol "r," is the most common correlation coefficient reported in
the biometeorological literature. It assumes or demands that either
interval or ratio scales are used. Although the actual calculation
is not difficult, the details are not required here.

The reader can attain a "gut feeling" for different correlation
coefficients by observing scattergrams of data for which r-values
have been calculated. In Figure 4, examples of correlation coefficients are given from very low values (r = 0.21) to very high values
(r = 0.98) approaching 1.0 (a straight line).

The general pattern is obvious: the greater the deviation from
a straight line or linear relationship between two variables, the lower
the correlation. Even a correlation coefficient of 0.72 contains a
great deal of scatter. The existence of correlations below r = ±0.4
are sometimes difficult to detect with the naked eye in a scatterplot.

To say that weather variable X is correlated with human variable Y is a trivial statement, since technically any r value that is
not zero indicates a correlation. A procedure must be applied to
determine whether a correlation is not just a random value. After
all, calculations with numbers by necessity generate numbers. They
do not indicate that the numbers mean something. How does one
separate real effects from random operations or chance?

The criterion used for meaningfulness is statistical significance.
When a correlation coefficient is statistically significant, it is
assumed that the contribution of chance to the value has been reduced
substantially. Depending upon the level of significance, the degree
to which the correlation may be due to chance can be stated. For
example, if a correlation coefficient is significant at the p (probability) = 0.05 level, the same value would occur 5 out of 100 times
by chance alone. With every 20 correlations calculated, one r-value
would be significant at $p \leq 0.05$ by chance alone.

If the significance level is p = 0.001 level, the probability that
the correlation were due to chance would be only 1 out of 1,000.
The acceptance of correlations with low p-values, such as p < 0.05,
increases the risk of including chance correlations, especially when
hundreds of correlations are calculated. The demand for high p-

values ($p < 10^{-6}$) would eliminate weak, and the most frequent bio-meteorological, correlations.

Two factors influence the level of a coefficient's significance: the size of the value and the number of pairs of observations involved. An r-value of 0.86 is more significant than an r-value of 0.46. The total number of observations contributes to the level of significance as well. The more observations (20, 40, or 100) involved in the correlation, the less likely the resulting coefficient is due to random fluctuations. Small groups of observations (such as 5 or 10) are prone to extremes that are influenced by a large pool of unspecified or chance factors.

The larger the number of observations involved in the calcula-tion of a coefficient, the lower the acceptable value of the coefficient for some statistical level. If one chooses a p-value < 0.01 as the acceptable level for any coefficient (that is, the probability is less than 1 out of 100 that it would occur by chance alone), lower and lower r-values become acceptable as the number of pairs of observa-tions increases. The minimum r-value acceptable with 10 pairs of observations is ± 0.77, with 20 pairs of observations is ± 0.56, with 50 pairs of observations is ± 0.36, and with 100 pairs of observations is ± 0.23.

Given this characteristic of statistics and correlations, one should be very wary of such statements as, "The correlation between suicide rates and changes in barometric pressure was statistically significant." One must know the numerical value of the correlation coefficient and the number of observations involved. With sufficient pairs of observations, even the correlation in Figure 4a can become statistically significant.

Let us reduce the statistical significance problem to a "close-to-home" example. Suppose the correlation between the dosage of aspirin and the possibility of death, ranging from a probability of almost 0.0 to nearly 1.0, displayed an r = 0.30. From the scatter-gram noted in Figure 4a, would you be willing to bet your life on a particular dosage? Suppose the correlation were r = 0.72, a really impressive correlation coefficient for scientific journals? Would you risk taking aspirin with this type of scatter between specific dose and total effect?

Practical Significance

Statistical significance does not assure practical significance. If correlation coefficients are so small that they are not detectable by observation or that they can be included within the rubric of hundreds of other background noises, their practical value for pre-dicting or understanding weather is limited.

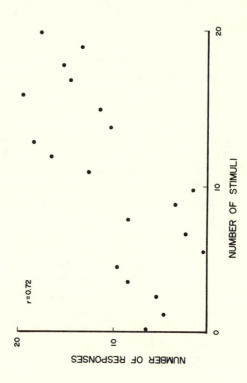

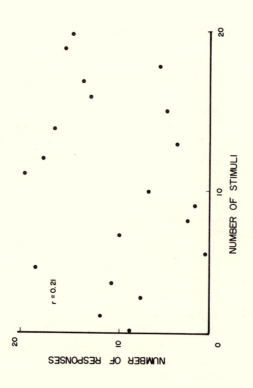

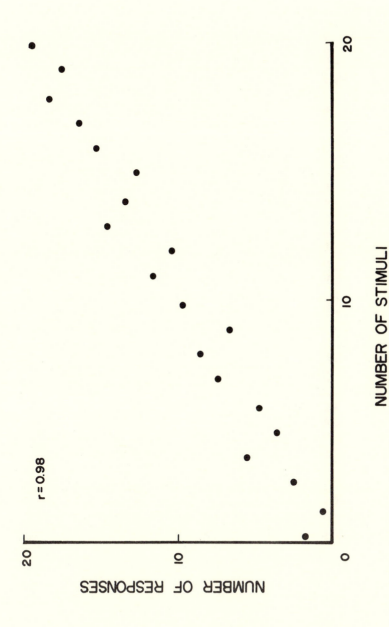

Figure 4   Scattergrams demonstrating various degrees of relationship between variable Y and variable X.   Pearson product-moment correlation coefficients (r) have been calculated for each of the scattergrams.

If a correlation coefficient between some human behavior or response and weather change were significant, one must ask, "What is the real potency of the weather effect?" How much of the variability of the behavior or response can be accommodated or explained by the weather change? Does weather account for a little or a lot of the behavior presumably affected by the weather change?

A large part of the answer to these questions involves a simple operation: taking the square of the correlation coefficient. This value, called r^2 (r-squared), specifies the percentage of the total variance in one measure that is explained by the other. Presumably if the r^2 were 100 percent or if the sum of different r^2s were 100 percent (under certain conditions), one could totally explain the variability in one measure as a function of the other and vice versa. Because of the internal variation in measurement procedures, the small but constant variation in instruments, and human participation, 100 percent accommodation of variance rarely happens.

Applied to weather-human response correlates, r^2 calculations place extreme claims in proper perspective. Suppose the correlation between the swelling of joints (in millimeters) and fall in barometric pressure (in millibars) were an impressive $r = 0.42$ ($p < 0.01$). (Correlations of this magnitude or less are used as rationales from which political policies are made, economic trends are boasted, and psychological treatments are given.) The r^2 of the coefficients is 0.42×0.42 or 0.17 (17 percent). This means that only 17 percent of the variance in joint swelling is accounted for by knowing the variation in barometric pressures. The remaining 83 percent of the variability in joint swelling is not specified by the correlation.

Even very respectable correlations, such as 0.80, lose their luster when the r^2 is taken. In this case, the r^2 is 0.8×0.8 or 64 percent. Despite the very high and significant correlation, only 64 percent of the variability in one variable can be accounted for by the other. The remaining 36 percent of the variability is due to other factors.

The majority of weather-human response correlations reported in scientific journals range from ± 0.30 to about ± 0.60. Surprisingly, this range of r-values includes not only vague complaints and mood changes but also more discrete events, including traffic accidents, heart failure, or death correlated with blatant weather changes, such as snowstorms and fog. With these correlations, one can expect to account for only from 10 percent (0.3×0.3) to 36 percent (0.6×0.6) of these weather-affected behaviors.

There is nothing magical about correlations around ± 0.30. In actual fact, correlations between weather changes and human responses range from $r = \pm 0.01$ to $r = \pm 0.60$. However, correlations less than 0.30 (on the average, depending upon the number of pairs of observations) would not be considered statistically significant. Consequently, because of the bias to publish only statistically significant results in scientific journals, these data would not be published.

With the exception of clear mechanical effects—for example, rain interfering with automobile tire friction—an antagonist to human-weather correlations might evoke a valid and damning question to all of biometeorology. If there are thousands of specific human responses and the level of correlational significance is $p \leq 0.05$, wouldn't there always be some human behaviors paired with weather changes by chance alone?

Correlation is not Causation

When a significant correlation is found between some restlessness measure and altered barometric pressure, there is a common erroneous tendency to conclude that the pressure change caused the behavior. Popular press and quasiscientific magazines frequently make claims that "Scientists prove barometric pressure causes headaches" or "Biometeorologists prove cold fronts cause heart attacks." The problem is not that simple.

A correlation coefficient is a statement of mathematical relationship between the variability in one measure and the variability in another measure. The coefficient does not tell the causative factor. When a significant correlation exists between heart failure and influxes of cold, high pressure air masses, no direction is implied. From a strictly mathematical point of view, it is just as likely that the heart attacks caused the cold front influxes as that the cold front influxes caused the heart attacks.

Direction of effect is usually determined according to some theory or to common sense; it is not given by the coefficient. When a correlation exists between suicide rates and lunar phase, people assume that the lunar phase is responsible for the suicide rates, not that the suicide rates are responsible for the lunar phase. However, this conclusion is an interpretation of the mathematical result. A laughing response to the conclusion that suicides cause lunar phase is a consequence of general belief, not objective analysis.

When significant correlations exist between human behavior and environmental events, the present postulates of science insist that the environmental events cause the behavior. It is antagonistic to our philosophical assumptions of humans and the universe that a significant correlation between weather and human behavior may also indicate that behavior influences weather. The correlation does not support either assumption; it only indicates that a relationship exists.

Considering this bias, one should carefully evaluate interpretations of correlations in which the direction of effect is influenced by basic philosophical issues. If a correlation between two variables with similar philosophical bases of cause and effect occurs, implicit associations could determine the reference variable. Whereas a chemist might view a correlation between barometric pressure and temperature as due to temperature causing changes in molecular numbers per unit area, a geologist might conclude that tectonic pressures generate different temperatures. Both conclusions are incorrect if made on the basis of only a significant correlation coefficient.

The Third Factor

Significant correlations between two measures do not eliminate the possibility of a third factor that causes or influences both. If this condition exists, the correlation between variable X and variable Y is due primarily to the fact that both X and Y are related to the third factor Z. Without the correlation of Y to Z and X to Z, the correlation between X and Y would not occur.

The third-factor problem is endemic to all correlational analyses. It indicates that the significant correlation between heart failure and barometric pressure may be due to a third factor that is correlated with both measures. It indicates that the relationship between suicide and lunar phase may be a consequence of a third factor with which both are correlated.

Selection of a favorite weather variable, such as barometric pressure, to correlate with human behavior without involving other common variables is a prelude to poor procedure. The mere occurrence of a significant relationship between barometric pressure and arthritic pain does not prove causation nor, for that matter, even a functional relationship.

Objectively, little can be stated about the explanatory variable without controlled experimental studies under strict laboratory conditions. With so many, many changing, highly interrelated variables in the weather matrix, the insistence that only one is responsible for a given behavior, merely because the two are correlated, occupies a special position between ignorance and stupidity.

The Problems of Causality

Causality is coupled almost entirely with the level of scientific discourse in which one is behaving. At the level of the atom, one can legitimately and logically conclude that atoms cause other atoms' behavior. At the level of physiology, one can conclude that a change in some physiological condition, such as cell-division rates, is responsible for the organ's activity. The major difficulty arises when the human thinker crosses discourse levels.

Crossing discourse levels is completed most frequently and inadvertently between scientific language and everyday language. Within the limits of language—which is a nominal scale for the most part—causality is a verbal label used to describe how one event is evoked by another event. If you strike a match and the match lights, you conclude that the strike caused the match to flame.

As long as the term "cause" is used within the very gross measurements of everyday language, few internal inconsistencies arise. However, as the resolution of detail becomes greater with the use of precise numerical scales that allow counting along a potentially infinite range, the contradictions of causality exacerbate. Variables begin to emerge, previously included under the same words in everyday language, that are quite different. They appear to contribute to the same events.

Technically speaking, the verbal statement "I caused the match to light when I struck it" is an arbitrary and superficial statement. Between the strike and the one millisecond before the flame ignites, billions of different events happen at molecular and atomic levels. These levels can be measured and differentiated, unlike the overly inclusive nature of human language. Which of the billion events actually caused the match to flame?

Statements of causality are helpful in everyday language to reduce the anxiety-provoking aspects of an unstructured environment. A highly unpredictable environment is a very aversive and anxious stimulus for the human being. Causality is frequently correct in situations in which the series of events leading to an observation occur very quickly or in which the series of events are homogeneous. In the match example, the events leading to the event occur in millionths of a second and the material (the match) is more or less the same material, that is, phosphorous and nitrates.

When these two conditions are less precise, causality becomes less applicable. In weather-human correlations, the time between the actual change and human response could be from minutes to hours. The factors evoking the human response could be many. Using a causality statement to describe weather effects upon human beings is similar to determining, at the molecular level, which of the many reactions occurring at the "same time" are responsible for some observable response.

FACTORS LIMITING CORRELATIONAL STUDIES

Correlational methods have several implicit or internal limitations that can create weather effects when there are none and ignore weather effects when they are present. A measurement tool or a procedure of analysis is limited by the competence and experience of the experimenters who use them. Some common problems encountered with correlational proofs of weather effects follow.

The Confounding Factor

A confounding factor is a third variable that is not involved in the correlation, but is responsible for its occurrence. Often confounding factors are intervening variables that occur by accident or are produced simultaneously by some other nonweather source. Whatever their source, confounding variables alter the display frequency of behaviors leading to the response in question.

The first example of a confounding variable was found in a correlation between barometric pressure and traffic accidents reported by a small community agency. The correlation indicated that more accidents occurred on days with low barometric pressure than on days with high barometric pressure. Of the 200 accidents in the town during the analysis period, 150 occurred during low pressure while only 50 occurred during high pressure. The conclusion, as you might expect, was that accidents are three times as likely to occur during low pressure days than during high pressure days. There was also an explanation for the relationship: "low pressure clogs the mind with fluids that slow it down."

However, there was a confounding factor: the number of cars on the road during high or low barometric pressure days. This small community had been involved in a physical fitness program. Members of the community were encouraged to ride bicycles or walk to work whenever possible, behaviors more likely to occur on sunny days characterized by high barometric pressure. When the data were reviewed, it was found with some embarrassment that an estimated 10,000 cars were on the roads during high pressure days, while an estimated 30,000 cars were on the roads during low pressure days. With three times as many cars on the roads during low pressure days, the likelihood of an accident was approximately three times as great.

A second example of a confounding variable was found in a study involving suicide incidence during influxes of warm weather. The report was a simple one: days of warm weather were associated with 83 percent of all suicides, while days of cold weather were asso-

ciated with the remaining 17 percent of suicides. According to the
report, one would expect that if chance were involved only 50 percent
of the suicides would occur during the warm periods, and the remaining 50 percent would occur during the cold periods. The conclusion:
warm weather produces suicides. Again, an exotic explanation was
presented.

But there was an unbelievably simple confounding factor that
should have been considered by someone on that great analysis team:
What percentage of the days during the analysis period were warm
weather days? A second investigation of the data clearly demonstrated
that during the analysis intervals 80 percent of the days were warm
(the city had a heat spell). More suicides occurred on warm days
merely because there were more warm days, hardly an insight.
The list is distressingly long. A few common confounding factors
of weather matrix correlates follow.

The full moon correlation with rapes is confounded by more
light by which to see victims and more people or potential victims
walking around during the full moon.

The correlation between low pressure areas during the wintertime and heart failures is confounded by snowfall. Snowfall is more
likely with this weather as are the mechanical consequences of clean-
ing sidewalks.

The correlation between thunderstorm activity and traffic
accidents in countries in which thunderstorm activity has a five to
one chance of occurring during the rush hours.

The correlation between arthritic complaints and cloudy days
when sufferers are more likely to stay inside their houses and engage
in tasks requiring dexterity, such as sewing, knitting, writing letters.

Single-Factor Correlation

Single-factor correlations have been discussed briefly. This
operation is very popular in "quick and dirty" (Q and D) studies,
especially when some private theory is involved. Single-factor
correlation involves only one stimulus from the weather matrix
correlated with only one human behavior. All other parts of the
matrix are ignored.

Claims about barometric pressure correlates with human
behavior and lunar phase effects with human pecularities are frequently based upon these analyses. The reader is not informed
about correlations with possible confounding variables, such as
rain- or snowfall, or about local weather variations correlated with
lunar phase, such as relative humidity or temperature, that are well

known to influence behavior—without recourse to exotic explanations. Compared with these local weather stimuli, lunar phase or barometric pressure may be very weak correlations.

Seasonal Effects

Seasonal effects are included coincidentally in weather-behavior analysis when great intervals of time are involved, such as two or more months. Pulp writers frequently fail to differentiate the corre- lates of weather from those of season. The statement, "More colds occur during the winter months," is coupled invariably with the statement, "Colds are more likely a few days after the influx of cold air."

Distinctions between seasonal and weather effects are important for both analytical and theoretical reasons. Seasonal effects involve slower fluctuations in the environmental parameters that are the baselines upon which transient short-term fluctuations (weather) exist. As a whole, seasonal effects, such as diseases occurring during the winter versus those during the summer, are much greater than weather effects.

When analyses are completed over the entire year, correlations between human behavior and temperature are contaminated by the predominance of low temperatures during the winter months and by the prevalence of warmer temperatures during the summer months. The statement, "Cold air is associated with the common cold," may not reflect a weather effect but, rather, a consequence of season.

SUGGESTED PROCEDURES FOR CORRELATIONAL STUDIES

Although correlational analyses between weather matrixes and human behaviors are complex and replete with confounding varia- bles at every step, the pitfalls of popular Q and D publications can be reduced by implementing a few procedures. These suggestions are not intended to be exhaustive, but they may be helpful in the initial attack against a mass of data.

Use the Weather Matrix

Do not correlate the behavior in question with just one or two stimuli from the weather matrix. Include obvious confounding factors, such as snowfall, rainfall and wind speed. If possible use temperature, humidity, barometric pressure, sunshine hours, wind speed and direction, geomagnetic activity, solar measures (X-ray bursts), and lunar phase.

Use Absolute and Rate of Change Measures

Correlate both absolute measures of the weather matrix variables as well as rate of change measures. For example, include mean barometric pressure, change in barometric pressure per 24 hours, and greatest change in barometric pressure within any 4-hour period during the 24 hours. Determine similar values for temperature and humidity and other hourly recorded measures. If possible, change measures should be preceded by their direction: + for increase and - for decrease. Most canned statistical programs will accept these data.

Correlate the Weather Matrix

Complete a correlation table or matrix between all measures of the weather matrix before adding the measures of human behavior. Determine which variables are highly intercorrelated. Do not choose a variable as an explanation for the human behavior just because it makes sense.

Use More Than One Behavioral Measure

A common flaw in correlational designs is to include only one behavioral measure, such as numbers of suicides or accidents. In light of problems with confounding factors, add the following, if possible, to the correlations: a measure of general activity of the population, such as numbers of cars on the roads (for traffic accidents) or numbers of total hours of operating industrial equipment per day (for accidents). If possible, include variables of other supposedly unrelated activities, such as daily fluctuations in movie attendance or calls to emergency services, in order to determine if some general factor contributed to the specific response in question.

Lag Correlations

Weather on the day of the behavior may not be the primary correlate of the behavior. In context of the homeostatic model,

weather changes the day or days before could be more highly correlated with the human response. If the human response requires
several hours or days to be elicited, daily correlations would obscure
any strong effect.

Lag correlations should be completed between daily human
behaviors and the weather variables from one to five days before.
As a control for possible intrinsic errors within the correlation or
the data, further lag correlations should be made between behavior
and weather conditions from one day to five days later.

Watch for Intrinsic Variability

The range of some weather factors, such as barometric pressure, does not change greatly over the year. Consequently, correlations with behavioral measures, which also do not change in range
significantly over the year, will not be hampered by excessive variability. The range in temperature, however, is another problem.
During the winter, for example, temperature may range from -10°C
to +10°C, while summer temperatures range from +20°C to +40°C.
The two ranges do not even overlap. Now suppose the behavioral
measure shows the same range for each season of the year. During
the winter, it ranges from 10 responses to 100 responses (admissions
or whatever), while, during the summer, it also ranges from 10 to
100 responses. Suppose that within the -10 to +10 range, the correlation with responses is 0.7 and that within the +20 to +40 range, the
correlation with responses is 0.7. If the entire temperature range
were used, from -10°C to +40°C, the relationship between temperature and response numbers would be masked.

Watch for Curvilinear Relationships

Simple correlations work on the assumption that the relationship
between variable X and Y is a straight line. X will increase or decrease in linear proportion to the change in Y. A curvilinear relationship is shown in Figure 5. If one calculated the actual correlation
coefficient, the relationship between relative humidity and respiratory complaints would not be significant. However, a clear inspection
of the data indicates that the two are not related in a straightline
fashion. As humidity increases, the numbers of lung complaints
decrease, but then, near 70 percent humidity, the numbers of lung
complaints increase once again.

Without at least visual inspection of the data, possible weather-
human behavior correlates can be masked. Plotted scattergrams

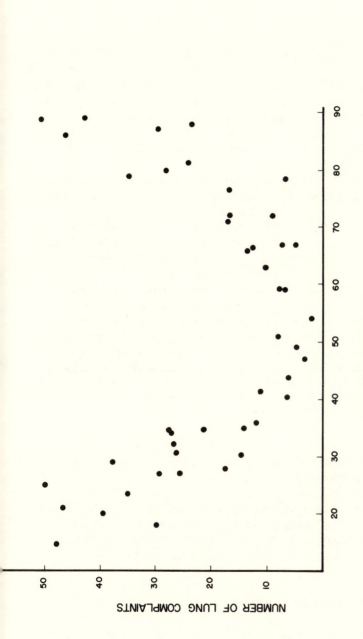

Figure 5   Scattergram demonstrating a nonlinear relationship between the number of lung complaints and the relative humidity of the air. Since most correlation equations assume a linear relationship within the analyses procedures, this strong curvilinear relationship would not be detected by simple observation of the correlation coefficient.

are essential since they allow a general view of the data pattern that would detect nonlinear profiles and since they quickly project any major errors in typing or punching or in improper analysis.

Watch for Weekend Effects

One of the greatest sources of variability within human behavior existing at time intervals relevant to weather effects is the weekend. During the weekend, great extremes in human experience are reported. If correlations are completed with behavioral data, especially when absolute numbers of responses are involved, partial out the effects of weekends.

A hypothetical example is presented in Figure 6. Suppose an apparently significant ($p < 0.001$) correlation ($r = +0.53$) exists between the number of traffic accidents and barometric pressure. Beware of weekend effects. Close inspection of the closed circles (data from weekends) demonstrates that the source of the significant correlation emerges from accidents on weekends. Without these data, the correlation is not statistically significant.

Weekends with high barometric pressure (sunny days) are associated with more cars on the roads, while weekends with low barometric pressure (cloudy, rainy days) are associated with fewer cars on the roads; people stay home. Whereas the number of cars on the roads during weekdays is fixed by common factors as job requirements, those during weekends have a much greater range.

Replicate

Always repeat your analysis to determine whether any significant correlation occurs within another set of data. Even highly significant correlations may occur by chance. One can check correlations within the data pool by dividing the pool randomly into two parts and computing correlations separately on the two parts. However, artifacts or confounding factors in the pool will be carried in both cases. Preferably, at least one other pool of data should be collected and analyzed to determine the reliability of any effect.

THE WEATHER-SENSITIVE PERSON

Failure to demonstrate weather effects, once the precautions and procedures mentioned have been followed, may reflect the operations of averaging. If a normal population were selected for study,

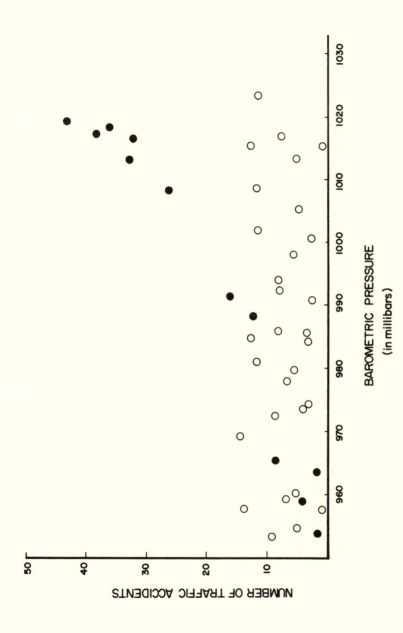

Figure 6   The relationship between the number of traffic accidents and barometric pressure. Open circles indicate data from days of the week, while closed circles indicate data from weekends.

a small percentage of people very sensitive to weather changes would not be detected. The contribution of their weather-sensitive behaviors to the large mass of measurements would be masked by the sheer weight of the analysis.

The Obscuring Average

Suppose a rating scale of mood, ranging in units from 1 (very, very depressed) to 10 (very euphoric), were used to indicate private experiences of mood. In order to maintain a representative sample, you choose 100 people randomly from the population. You calculate the average of the mood ratings on each day for one month and find that the mean (average) score is 5.5, well within the expected value of variation. Apparently, you conclude, weather does not influence this human behavior significantly.

If 10 percent of the population had been very weather-sensitive, the average would not have demonstrated them. Suppose 90 of the 100 subjects scored between 4 and 6 (with a mean of 5) and the remaining 10 of the 100 scored an extreme value of 10 (a maximum effect). The weighted mean $90 \times 5.0 + 10 \times 10$ would be equal only to 5.5. The ten weather-sensitive people would have been masked by the majority of nonresponsive subjects.

Studies Based upon Complaints

One of the most comprehensive statistical studies of the tendency to report weather sensitivity was conducted by Volker Faust and his colleagues in Basel, Switzerland (Faust, 1976; Faust, Weidmann, & Wehner, 1974; Faust & Sarreither, 1975). These experimenters collected questionnaire data from a number of different populations that included 1,600 students, from 13- to 20-years-old; 800 healthy normal adults; 380 diagnosed schizophrenics; 160 diagnosed depressives; 100 diagnosed neurotics; and several small clusters of other diagnostic categories.

According to their results, between one out of every four and one out of every two people reports weather sensitivity. Whereas only one out of three or four people in the normal population reported sensitivity, one out of two people diagnosed as neurotic or depressive reported sensitivity. People in diagnostic categories that display problems communicating and labeling their private experiences (for instance, thoughts, internal states, or gut reactions), such as schizophrenic patients, reported little weather sensitivity in this study.

Although weather sensitivity appeared to occur at all ages, from infancy to late adulthood, clear peaks were noted. Adolescents and adults between the ages of 40 and 50 showed marked increases in the report of weather-related symptoms. After 60 years of age, the relative number of reports appears to become constant. Females reported more weather sensitivity than males. Both age and sex differences were characterized by altered emphases upon symptoms clusters, rather than upon different symptoms.

Very weather-sensitive people, especially depressive and neurotic females, said they could feel weather changes three or more days before the changes occurred. Males diagnosed within similar categories considered their sensitivity to range only a few hours before the weather change.

Limits of the Term

Weather sensitivity has been used implicitly as a nominal scale. In the past, people have been considered either sensitive or not sensitive. Clearly, this simplistic dichotomy, especially if loose criteria were used to define sensitivity, is an exercise in error. Human beings who are sensitive once a year would be grouped with people who suffer weekly.

Classic techniques to determine weather sensitivity are contaminated heavily by verbal complaints. If a person complains a great deal and blames it on the weather, he may be considered weather-sensitive. Psychologists are still measuring the daily variations in human mood, pain complaints, and general experiences that would occur anyway, regardless of weather changes, due to the internal makeup of the organism. Paired association with weather may be only superstitious conditioning.

Theoretical Bases of Weather Sensitivity

Assuming the autonomic nervous system is a primary correlate of alterations in the weather matrix, some basic statements can be made concerning weather sensitivity. Since there are two components of the autonomic nervous system—sympathetic and parasympathetic—each with different response profiles and two (simplistic) kinetic options—stable or labile (unstable)—an elementary model can be postulated to predict weather sensitivity.

A weather-responsive person should be sensitive to subtle environmental events, especially of the matrix or diffuse type, and autonomically unstable. This lability would foster periods of excessive excursion, such as overshoots or undershoots, around stable conditions, for from hours to days, in response to distant or local weather conditions. The excursions would be experienced statistically depending upon autonomic noise from psychological factors.

Greatest sensitivity to weather changes should be reported by people who are labile and parasympathetic dominant. The basis for this sensitivity over sympathetic dominance is contingent not upon some special property of the parasympathetic system, but rather upon the phenomenon of contrast enhancement.

Daily, waking activities are characteristic of sympathetic behaviors in general. Response to massive sensory stimulation, complex decision making, and discriminate thinking are dependent in general upon the activity of sympathetic arousal. Parasympathetic activity during waking periods is normally counterproductive to these higher processes of social and cultural behavior.

A person who is already autonomically labile and parasympathetic dominant would have difficulty dealing with situations involving a great deal of thinking or complex, fast decision making, such as driving in rush-hour traffic. He would be very prone to drifting into a semiactive and relaxed profile, especially in situations that breed boredom: operating a drill press over and over again or driving upon the same old lonely road for hours.

Stimulation from weather changes would push such a person over the edge. Weather change would be the precipitating factor to induce the necessary increment of parasympathetic interference with complex waking behavior. The person would be more sluggish, report more pain (from sources not detected during sympathetic arousal), and be more responsive to the whole-body image. Physiologically, the blood pressure would be lower, resulting in a greater likelihood of fainting, black spots in the visual field, or orthostatic hypotension (faint feeling when you suddenly stand up).

A labile, sympathetic dominant person would be facilitated by weather change during the day. If weather change further stimulated sympathetic profiles, this person would report greater energy, more strength, faster thinking, an elevated mood tone (perhaps even to euphoria), and an enhanced self-image (when life feels really great and nothing, absolutely nothing, goes wrong). Physiologically, this person would show faster heart rates and higher blood pressure.

During the night when the person becomes relatively parasympathetic dominant, weather changes would adversely affect the sympathetic, labile person to a more significant degree. Weather stimulation during nighttime body changes would enhance greater sympathetic activity, which is not conducive to sleeping behaviors. This person would report more difficulty going to sleep and staying asleep. She would awaken many times during the night.

Since the sleeping portion of human behavior is not associated with events linked with accident probabilities—most people don't work drill presses in bed—or with behaviors conducive to private experiences—thinking per se is rarely displayed during sleep—the types of behavior associated with weather change will be less frequent.

Percentage of Weather-Sensitive People

If this mode is correct, labile sympathetic and parasympathetic dominant people should comprise the category of weather-sensitive. According to Gellhorn and Loofbourrow (1963, pp. 224-53), 8 percent of the normal population is parasympathetic dominant while 8 percent is sympathetic dominant. The remaining people are mixed types.

From their values, approximately 16 percent of the population or about one out of very five or six people should demonstrate physiological weather sensitivity. Since half the number of weather-sensitive candidates would show their effects during the night, clear weather associations should be evident in only 8 percent of the population or about one out of ten people. These values are too small to be examined through random sampling or gross averaging. Pre-screening of subjects is mandatory to demonstrate this form of weather sensitivity.

With respect to specific groups or categories of people that are responsive to weather change, one would predict any condition enhancing autonomic lability to facilitate physiological sensitivity. Females in general would be more sensitive to weather changes than males, especially in light of the autonomic coupling with hormonal periods. Life periods associated with increased physiological/psychological stress, such as adolescence and the fourth and fifth decades, should increase the probability of weather responses. Psychiatric populations, including psychotic categories, even though they may not report weather sensitivity, would demonstrate behavioral alterations correlated with the weather matrix.

PHASES OF THE WEATHER MATRIX

Although the air mass can be conceptualized and individual candidates of the weather matrix can be measured, a third dimension of analyses, weather phase, may be helpful in determining the existence of biometeorological effects. Changes in the weather matrix occur systematically over several days, with semialternating sequences of clear, cloudy, stormy, and clear days. The concept of weather phase allows an appropriate time component between the relative stability of an air mass and the simple daily fragmentation into component stimuli.

Examples

Several authors have described phase diagrams of weather. One system was devised by Ungeheuer and reported by Tromp (1963). According to this scheme, there are six phases of weather (in the Bavarian region of Germany): average nice weather, very nice weather, extremely nice weather with slight foehn (wind) conditions, initial weather change, complete weather change, and initial weather improvement.

During Phase 1, the usual daily variations in temperature and humidity are associated with a partly cloudy sky. The barometric pressure is relatively high and stable. People report experiences of a cool, dry, refreshing effect. As the weather progresses into Phase 2, cloudiness decreases, barometric pressure decreases slightly, and rhythmic changes in temperature and barometric pressure become pronounced. People might describe it as pleasantly warm.

Phase 3, the extremely nice weather category, is typified by a few lenticular clouds and the first cirrus clouds, coupled with a continued and sometimes accelerated decrease in barometric pressure. The temperature rises as a result of increased solar radiation and the influx of warm dry air. The weather is usually described as unpleasantly warm or muggy.

Phases 4 and 5 are associated with changes in a variety of meteorologic stimuli. During the initial weather change, a rapid increase of clouds or thunderstorms is noted. The barometric pressure continues to display a steep fall. Temperature remains stable, but humidity begins to increase. The weather is described as sultry and oppressive.

During the complete weather change, a sudden influx of cold air under a heavily cloudy sky is associated with a cold front and rain showers. However, after the front passes within hours, barometric pressure begins to rise, temperature falls steeply, and humidity remains relatively high. People report an uncomfortable penetrating cold.

As Phase 6 predominates, the sky becomes less cloudy and blue patches appear intermittently. The temperature is still low, but rises as the humidity gradually decreases during the day. The weather is described as pleasantly stimulating but cold.

A second example of a weather-phase system is shown in Figure 7. As does Ungeheuer's scheme, the scale assumes a con-

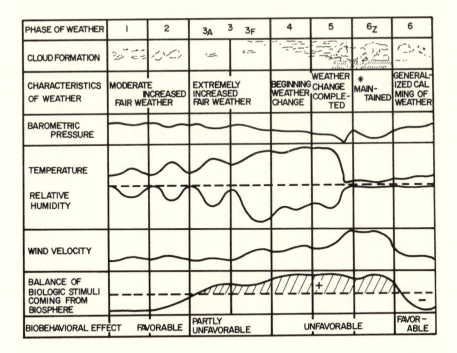

PHASE OF WEATHER	I	2	3A	3	3F	4	5	6Z	6
CLOUD FORMATION									
CHARACTERISTICS OF WEATHER	MODERATE INCREASED FAIR WEATHER		EXTREMELY INCREASED FAIR WEATHER			BEGINNING WEATHER CHANGE	WEATHER CHANGE COMPLE-TED	* MAIN-TAINED	GENERAL-IZED CAL MING OF WEATHER
BAROMETRIC PRESSURE									
TEMPERATURE									
RELATIVE HUMIDITY									
WIND VELOCITY									
BALANCE OF BIOLOGIC STIMULI COMING FROM BIOSPHERE									
BIOBEHAVIORAL EFFECT	FAVORABLE		PARTLY UNFAVORABLE			UNFAVORABLE			FAVOR-ABLE

Figure 7   Diagram of major phases of general weather systems,
demonstrating typical directions of meteorological
variables and possible biobehavioral consequences.

tinuum that can be divided into six major divisions. Different localities no doubt display other more appropriate phases. These should
be determined by careful analyses of the weather history within the
research area.

Different phases can exist for various periods of time, depending upon the overall or macroscale weather condition, and can be
influenced by the complex interaction of different air masses thousands
of kilometers away. Some of the phases can be extended in time
due to special or occasional conditions. For example, Phase 3F
in Figure 7 would represent a Chinook (hot dry wind) in the Canadian
Rockies, while Phase 6Z may indicate the abnormal continuation of
unstable weather due to the presence of an active cold front.

The three stimuli of temperature, relative humidity, and wind
velocity are self-explanatory: upward movements indicate increased
values of these measures. The categories of balance of biological

stimuli coming from the biosphere and biobehavioral effects are derived from physiological models and from meteorological mythology, respectively.

Importance of Weather Phases

All disciplines in science contain phenomena that occupy a unique position within the central concepts. These phenomena can be measured or conceptualized best with the space/time assumptions of the discipline. The molecule, for example, is difficult to conceive using merely the basic spatial and temporal assumptions of the atomic nucleus. Similarly, the air mass is almost impossible to describe without the integration of large areas and protracted time.

The use of the weather-phase concept allows the optimal selection of the time component that represents the basic unit of weather. Simple fragmentation of the weather matrix into components is important experimentally but may exclude temporal complexes of stimuli that are optimally biologically effective.

The common practice of daily measurement is also limited. Weather phases can exist for various durations, often exceeding a day and sometimes a week. If optimal weather stimuli exist within the temporal framework of from two to five days, analysis of human behavior on a daily basis would mask the actual impact of weather upon living systems.

Under the assumption of weather phases, human behaviors— from complaints to mortality—should not be analyzed according to fixed daily or weekly units. Instead, they should be analyzed according to the time-varying weather phase in which they occurred. Any disparity between the different numbers of days within different weather phases could be accommodated by a number of simple mathematical operations.

The serious consideration of flexible phase intervals, rather than fixed intervals or specific-stimulus analyses, is a logical extrapolation of the Lebowitz data (1973a; see Chapter 3). He noted that most meteorologically related human behaviors are interrelated within two- to four-day periods. In other words, the optimal analysis interval for summing or averaging the data would be from two to four days on the average, rather than a single daily unit.

This interval is quite variable and presumably related to interrelations in weather conditions. A number of individual stimuli of the weather matrix are interrelated within that time span. If one correlates temperature across days, the highest correlations occur at increments within plus or minus two days. A similar relationship occurs even for variations in other stimuli, such as geomagnetic storm measures.

Advantages and Disadvantages

The advantages of using weather phases rather than individual or group analyses should be evident. Phase analyses allow incorporation of the elusive and variable time component in order to maximally enhance the biobehavioral effect. Latencies of responses in people often require a stimulus of more than one day in order to evoke any change. For complete saturation to occur, several days of protracted presentation may be required as well.

We have found that dividing weather into six or seven types allows both an easier conceptual grasp of the complexities involved and a simple, first-order framework of analyses. Weather-related human behaviors, such as mining accidents, that have been hidden by daily analyses can be enhanced readily when analyses are completed as a function of weather phase.

The potency of the method can be emphasized by a simple example. Suppose the daily numbers of accidents range between 0 and 10. For days that are truly independent, the numbers of accidents on adjacent days should be a function of statistical population factors. Adjacent days that are controlled by some interdependent variable, such as a weather condition, should be biased in the same direction.

Suppose four days (Phase A) have accident numbers of 4, 0, 8, and 5, while another four days associated with labile weather (Phase B) have numbers of 7, 8, 6, and 7. Individual-day analyses would not show anything unusual about the eight days. They all fall within the normal 0 to 10 range. However, analyses according to four-day intervals indicate that 17 accidents occurred during weather Phase A while 28 occurred during weather Phase B.

Since the number of days displaying a particular weather phase can be counted easily, the magnitude of any human effect can be more easily correlated and displayed. One would expect contrast enhancement of human phenomena as a function of the duration of some phases. The positive feelings associated with a bright sunny day should occur most frequently on the first sunny day following a week of continual cloudiness than on the third sunny day or the first sunny day following only one day of cloudiness.

The practical advantage of only six or seven basic types of weather phases for a given locality can be seen in the massive reduction of data arrays. Whereas daily analyses of 100 days would involve 100 measures times the number of specific meteorologic stimuli involved (for example 100×10 stimuli, 1,000 groups of numbers), phase analyses reduce the number of observations to about 30 or 40 per 100 days. With 30 or 40 cases, one can quickly view any oddities or extremities in the data associated with a particular phase.

The primary disadvantage of the weather-phase concept is its dependence upon a complex criteria. The diagnosis of a particular phase would be dependent upon human discrimination or some complex equation with variable weighting for different weather stimuli. Since some areas are prone to quick changes in cloud formation or other meteorologic stimuli within a day, more than a single daily appraisal of the environment would be required. These problems are not insurmountable.

Surprisingly, few researchers have used a weather-phase analyses approach. Most of the weather-phase data have been published from the Bavarian region of Germany and some eastern European countries within which persistent patterns of weather have been correlated with human behavior. The approach contains comparatively few methodological problems compared with the results gained. However, this analysis requires a planned research protocol and a clear understanding of statistics.

6

TEMPERATURE, HUMIDITY, AND RELATED CONDITIONS

Temperature is a critical variable in most chemical reactions. In a parcel of air, temperature values determine the perceived characteristics of the atmosphere. The amount of water vapor that can be maintained in the air before condensation (precipitation) occurs is directly related to temperature. Other meteorological phenomena, such as wind and barometric pressure, are functionally coupled to temperature.

Psychological and physiological evaluations of air temperature pivot around the set points of the human homeostat. An average naked person would consider temperatures below 20°C to be cold and those above 30°C to be hot. These judgments could be modified significantly depending upon the amount of water vapor in the air, from near zero to near saturation, such as rain. If the person were outside, the evaluation of cold or hot would be influenced as well by the amount of sunshine or the presence of clouds.

All of these variables contribute to the general biological phenomenon of heat loss. Air temperatures below between 30°C and 37°C contribute to heat loss from the living system. Air temperatures above from 30°C to 37°C interfere with heat loss. Since the human being as a phenomenon must maintain a very narrow temperature band, stimuli from the weather matrix above or below these values can elicit significant and diffuse changes throughout the body.

TEMPERATURE: THE PERSPECTIVE

A fundamental assumption within the human concept of the environment is the existence of energy. Energy can be defined as the capacity to produce motion within space. Energy cannot be destroyed according to present concepts, but can be transformed into different forms.

Classic forms of energy are distinguished according to potential manifestation or according to actual (kinetic) display. Energy is also designated as a function of space. Atomic energy is the form that maintains the small space of the atomic nucleus, while electric energy is the form that maintains the motion of electrons within a larger space.

Heat energy is the capacity to produce motion at the level of the atom as a unit. What we call temperature is an arbitrary number for the relative activity of atoms. Atoms that move less frequently or are less active than another group of atoms are assumed to have lower temperatures. Consequently, the atoms within ice move more slowly within space than the atoms within tropical air. Presumably, if all energy were removed from a parcel of atoms, motion would cease.

Heat energy contains not only a dimension of molecular or atomic activity, but also one of quantity or potency. From everyday experience, we realize that a large pan of boiling water has the capacity to melt more snow than a teacup of boiling water. The measure of the quantity of heat is the calorie. By definition, one calorie of heat energy has the capacity to elevate the temperature of 1 cubic centimeter (cc) of water $1°C$.

The heat density or concentration of energy is related to the density of the matter involved. A group of molecules may have temperatures or activities comparable to thousands of degrees C. However, if only a few molecules were present in a given space, such as the low matter density of outerspace, the amount of total energy may be negligible. One could still freeze to death within a rarified mixture of gases within temperatures thousands of degrees C.

Since the concentration of atoms is relatively constant within the biosphere, the latter problem is not critical. Temperature measures in different regions of the biosphere would reflect, in general, the different amounts of heat energy available in the air masses. Caloric measures become important when one deals with interfaces between two matter densities, such as a person lying on a slab of metal or immersed within a body of water rather than exposed to normal air. In these situations, one must know the caloric capability as well as the temperature of the exposure medium.

Measurement of Temperature

The presumed activity in atoms is commonly measured on Fahrenheit and Centigrade temperature scales. Both measures implicitly use humans as a reference since the origin (0°) of both scales occurs within life temperatures. These scales are interval scales; they do not contain an absolute zero. Zero degrees on the Centigrade regime is the temperature at which water freezes, while 0 degrees on the Fahrenheit scale is 32 degrees below the freezing point of water. Both scales have the capacity for negative values.

The Kelvin scale of temperature measurement is a true ratio scale since it contains an absolute zero. Absolute 0°K is equivalent to -273°C and is the temperature at which atomic motion would terminate. This continuum is used also for designating heat energy emitted by stars and other large conglomerations of energy and matter.

However, for practical reasons, the Fahrenheit and Centigrade scales are used for organismic studies. Since life systems exist within a relatively narrow band of heat energy compared with the possible range that exists within the universe, the Kelvin scale is excessive. The selection of °F or °C is usually a matter of learning history. For convenience, the Centigrade scale is used in many studies. Fewer coefficients of ten (0, 10, 20, 30) are required to demonstrate the usual environmental ranges than with the Fahrenheit scale (30, 40, 50, 60, 70, 80, 90). For conversion: F = 9/5 C + 32, while C = 5/9 (F - 32).

Biological Significance

Heat or one of its various manifestations is the primary form by which life systems are maintained. Indeed, many of the inorganic processes within the earth, such as the formation of minerals and gases, involve the display of this energy form. Heat energy is released from foods to maintain the internal dynamics and biochemical reactions of the body. Heat energy is released externally, by various forms of combustion, to reduce the discrepancy between the ambient and biological environments.

The fundamental contribution of heat energy to the existence of mammalian systems can be seen in both the narrow temperature band in which they occur and the large portions of the biological structures oriented toward maintaining thermostasis. Whereas the biological temperatures of amphibians and less complicated organisms are subject to the potentially incapacitating effects of ambient temperature changes, mammals contain extraordinarily complex mechanisms throughout the body volume with which to maintain a very narrow temperature range.

Although skin temperature can deviate significantly around average values, the core temperature within the body rarely fluctuates by more than 1°F. If brain temperature drops from 37.1°C to 37.0°C—only 0.1°C—heat production shifts from a baseline of 20 calories per second to 40 calories per second. If the brain temperature drops to 36.8°C, the heat production increases quickly to 60 calories per second. If the brain temperature increases from 37.1°C—almost no heat loss—to 37.2°C, heat loss increases to 40 calories per second. By core temperatures of 37.4°C, the heat loss is about 70 calories per second (Guyton, 1971).

The narrow band of core temperature can be maintained in human subjects exposed to a surprisingly wide range of ambient temperatures. Naked volunteers exposed for several hours to ambient temperatures between 50°F and 110°F display no discernible change in core temperatures. When exposure temperatures are decreased from 40°F to 30°F, the average body temperature (0.7 × core temperature + 0.3 × skin temperature) drops quickly from 97°F to 70°F. When the exposure temperatures are increased from 130°F to 150°F, the body temperature raises from 97°F to 100°F. (These are general values and are subject to small but critical individual variations.)

Below and above the 37°C increment, complex behavioral patterns, such as thinking, concentration, and serial response sequences, are disrupted or suppressed. The person reports periods of drifting (intervals of time for which there is no or little recollection), inability to think, and, sometimes, a feeling of general loss of control. At more extreme deviations from the 37°C degree set point, the person may experience panic, intense anxiety, and a general intensely negative tone. Some people report enhanced feelings of suicide and self-destructive tendencies.

Below core temperatures of 30°C, human patients no longer display any form of temperature regulation. While between 34°C and 30°C subjects can still maintain elementary function, vital systems begin to deteriorate below this temperature. The hardiest subjects in concentration camps died around core temperature of 25°C, even if they had been slowly lowered to these values. Exposure to ice water produces this core temperature and death within about 30 minutes.

Above 38°C, the average person displays periods of distress, depending upon the origin and duration of the hyperthermia. Heavy exercise can raise the core temperature to 40°C, while fever therapy has pushed core values to about 42°C. Temperature regulation in general is seriously impaired above 40°C. These values are frequently associated with heat stroke and brain lesions. Above 43°C, few patients survive.

Considering the narrow range at which the human being optimally survives, one should not be surprised at the elaborate nature of response systems involved with maintaining this temperature.

The study of heat loss and heat gain, in fact, constitutes the major emphasis of environmental physiology (see Folk, 1974). Compared with temperature, other components of the weather matrix may seem trivial.

Simple Heat Physics and Physiology

Heat loss can occur due to at least four physical mechanisms: conduction, evaporation, radiation, and convection. Conduction is the movement (flow) of heat energy from areas of higher concentration to lower concentration within an unequally heated substance. Convection is a similar process in gases or liquids due to processes (currents) that carry heat away. Whereas evaporation absorbs heat from the surface, condensation replaces heat to the surface. Radiation is the transfer of heat, as electromagnetic waves, through space without warming the intermediate environment.

According to Folk (1974), a person with a core temperature of 37°C and a skin temperature of 32°C sitting in a room with an air temperature of 24°C will lose heat primarily through radiation, evaporation, and convection. Evaporative losses will contribute to about 23 percent of the heat loss, while convection into the surrounding air will average about 10 percent. The remaining heat loss (67 percent) will be partitioned into the from 5- to 20-micron (infrared) radiation emitted as radiant energy from the skin.

The mode of heat loss can shift from radiative to convective processes. With little or no wind, according to Guyton (1971), only about 15 percent of the total heat lost is through convection while, at 4 and 8 miles per hour (mph), 40 percent and 60 percent of the heat loss is through this method, respectively. If the person were directly exposed to a very cold medium other than air, such as water or a solid object, the rate of loss can be extraordinarily quick.

Heat physics is important for determining the rate and characteristics of thermal loss from the body boundaries. Radiative loss from the skin will increase as the discrepancy between the skin temperature and ambient air temperature increases. Similarly, the amount of evaporative cooling increases as the rate of fluid evaporation on the skin increases. Since the air immediately around the skin is approximately saturated with the skin fluid, the amount of evaporation will be influenced by the relative humidity of the air and the frequency of air displacement.

Heat physiology involves the biological mechanisms by which heat energy is delivered to the skin and is generated within the core. Factors increasing heat production include exercise, shivering, tensing of muscles (even imperceptibly), chemical increase of metabolic rate (release of adrenalin, for example), and the dynamic action of food during its conversion from raw material into the energy molecule adenosine triphosphate (ATP). Whereas about 44 percent of the energy from the breakdown of glucose is stored within this molecular form, the remaining 56 percent is released as heat energy within the body.

Heat loss can be decreased by shifting the blood flow toward the core of the body. This process usually involves peripheral vasoconstriction and central (heart and brain) vasodilation, a primary sympathetic nervous system response. Heat loss can be increased by peripheral vasodilation and by shifting blood flow to the periphery, thus exposing the warmer blood to the cooling effects from the skin. Sweating, increased air movement, as well as increased radiative surface (removing the clothes) facilitate heat loss.

The capacity for the human system to respond to temperatures below or above core levels is clearly lopsided. Whereas excess heat loss can be accommodated by a variety of different physiological processes, from short-term shivering to long-term endocrine activation, insufficient heat loss (in high ambient temperatures) is limited to evaporative cooling.

No doubt this discrepancy is tied to the nature of heat energy. In general, the release of heat energy into the environment is much easier than the removal (the physics are not complicated, but are outside the scope of this text). If a preferred living temperature were 20°C, a decrease of temperature to 0°C can easily be accommodated by a multitude of heat-generating conditions, including simple ambulation or escape. On the other hand, a 20°C increase in temperature to 40°C is more difficult to accommodate, biologically and technologically.

Temperature Plus Time Problems

Although temperature may appear as a simple, homogeneous concept, the biobehavioral impact of temperature changes is quite complex. Popular literature frequently combines and subsumes the daily, weekly, monthly, and yearly variations in temperature within the same category of potential biological effects. This tendency has produced a great deal of confusion and error concerning the qualitative and quantitative effects of temperature changes upon biobehavioral systems.

The combination of temperature and time forms potentially hundreds of different, discernible types of temperature stimuli. First, there are the slowly varying, large amplitude changes in temperature associated with seasonal periods. Upon these large variations, secondary daily and weekly variations are superimposed; these temperature stimuli are classically associated with weather. The third complicating factor is the rate of temperature change within any of the time scales considered: day, month, or even year.

Indiscriminate combination of different temporal increments of temperature change can lead to error. A common report in the noncritical literature is that low ambient temperatures cause death by stroke or cardiovascular disease. Indeed, the relationship between low temperatures (especially below -10°C) and death has been known for at least 200 years. Crude general application of statistics to monthly temperature averages and cardiovascular mortality clearly produce correlation coefficients between -0.8 and -0.95 in most countries!

Such correlations, however, are not a consequence of temperature within the weather matrix per se. Instead, they are indicative of much larger changes across seasons. Considering the known differences in biobehavioral reactions to stimuli changing over months compared with stimuli changing over days, one would expect the involvement of different mechanisms. Simple combination of changes in temperature associated with weather and with season could easily mask fundamental differences in the biological modes of effect.

Figure 8 indicates the three theoretical possibilities of the relationship between seasonal variations and daily or weekly variations in temperature. In Figure 8A, daily variations in temperature are commensurate with seasonal variations. Over the entire year, lower temperatures are associated with more deaths. Similarly, within each season, lower temperatures are associated with deaths. For this condition, the slope of the change or the angle of the line through both the weekly and seasonal changes is similar.

Figure 8B demonstrates the disassociative possibility. Whereas mortality increases with decreases in seasonal temperature, mortality is not related to temperature within the season. While the slope of the regression line remains the same for seasonal changes over the year, the line within the season is flat, indicating no systematic relationship to temperature. A crude correlation calculation would still show coefficients in the order of -0.80 due to the magnitude of seasonal variations.

Figure 8C suggests a less distinct and more likely relationship between mortality measures and seasonal versus weekly temperature variations. The slope of the regression line maintains the distinct relationship between seasonal temperatures and mortality. However, daily and weekly variations associated with temperature produce a similar slope that is less intense.

The distinction among the three conditions is by no means trivial. Different mechanisms by which temperature may influence

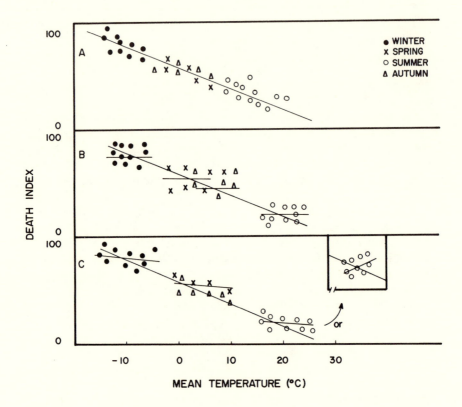

Figure 8   Various patterns by which yearly or weekly variations
            in temperature could contribute to a mortality measure.
            Both a continuous and a noncontinuous relationship be-
            tween temperature changes as a function of duration
            (weeks versus months) are shown.

the response measure are implied. Since the slope of the seasonal
and weekly correlations are similar in the first instance, the relationship between temperature and the human measure is simple and
linear. In the second situation, the effect appears to require a
threshold value (intensity or duration) since only seasonal changes
in temperature significantly influence the measure.

    The third condition, a modification of the first, suggests a
nonlinear relationship between temperature and the measure. Although weekly temperature values are correlated with the measure,
seasonal values are even more effective. Presumably a seasonal
temperature increment twice the value of a weekly one (that is, a

difference of 5°C versus 10°C would influence the measure by more than a factor of just two.

The interaction of seasonal and weekly temperature variations can be resolved by considering both the law of initial values and the systems models. The law of initial values indicates that the intensity of the effect will be influenced by the baseline or ambient temperature at the time of the stimulus change. If the ambient temperature has already pushed the organism into instability or beyond the limits of response, small fluctuations in temperature may not be very effective. One must consider both baseline (seasonal variations) and short-term transients (weather-related changes) superimposed upon the baseline.

The possible combinations of contradictory changes that could be associated with temperature do not terminate here. Whereas one can easily comprehend a yearly variation in temperature-related diseases associated with winter low temperatures and summer high temperatures (in the northern hemisphere), other factors can modify this pattern for some diseases. The change in temperature per month is not constant across the year. Some months show greater changes in temperature even though they may occupy intermediate absolute temperatures.

This problem can be seen in Figure 9. The single dotted-line curve shows the mean temperatures for each month, averaged over a 100-year period for the city of Philadelphia. As expected, the lowest average temperatures occur during the winter months and the highest temperatures occur during the summer months. The total range from average minimum to average maximum is about 45°F.

However, note that the months that display the greatest rate of change occur during the rising and falling parts of the cycle. March, April, May, October, and November show the greatest change in temperatures. Even though these months have intermediate average temperatures, the change in temperature exceeds those of the coldest and warmest months. Diseases that are rate-dependent— that is, perturbations that are contingent upon change rather than upon absolute values—would be expected most frequently during these periods and not during the more stable months of the midwinter or midsummer.

One could further add a second-order time component by noting which months display the greatest rate of the rate of change. From Figure 9, such changes are greatest between February and March and between November and December. If one superimposed the weekly and daily variations of temperature associated with the continual inundations from the weather matrix in conjunction with the varied and frequently unspecified shape (ramp, impulse, step) of the

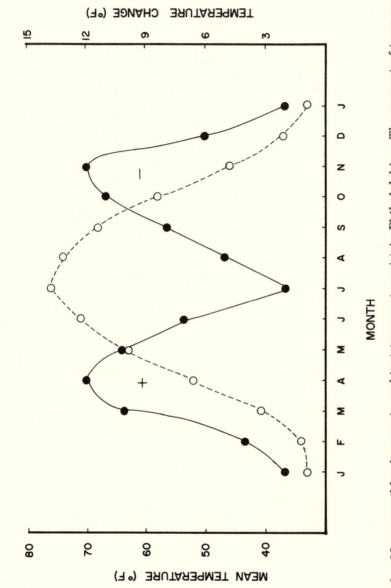

Figure 9   Mean monthly changes in ambient temperature (o) in Philadelphia. The amount of temperature change per month is demonstrated on the right Y axis and is indicated by closed circles (●).

<u>Source:</u> Modified from Boles and Westerman (1954).

temperature change, the critical thinker quickly realizes that temperature effects, in theory, are not simple.

SEASONAL TEMPERATURE

Several methodologically sound studies have demonstrated that yearly variations in temperature can account for the majority of the variability in death rates within several large northern hemisphere cities. Figure 10 is a modification and reproduction of the essential results of Sakamoto and Katayama (1971). Clearly, the major periodicity in death rates from heart failure is associated with variability in seasonal temperatures. Once this variability is partialled out, only irregular variations remain. From this study, one would expect that weather would have a very slight influence on these death rates and would contribute, at most, to the small irregular variations that remain after the major analyses.

Human Behavior and Seasonal Temperature Variations

A number of different behaviors, organ conditions, and body chemical changes have been linked to or correlated with annual fluctuations. Some of these periodic human behaviors are shown in Table 1. These data are derived from gross visual inspection of annual and monthly curves. The nonbehavioral measurements have been selected from Tromp (1963) and are intended to be illustrative rather than instructional.

Many of the conditions mentioned in Table 1 have been coupled theoretically or indirectly with seasonal changes in ambient temperature. However, few studies using statistically clear methods have clearly demonstrated that monthly or seasonal variations in temperature can actually account for variability in these diseases or behaviors. The common procedure of merely overlapping monthly averages of temperatures over the year with monthly behavior or disease frequencies is not sufficient proof of cause and effect.

Many biometeorological studies have involved monthly averages or totals of mortality measures and monthly averages of temperature for different geographical localities. These data are easily accessible since they are published routinely by most governments. The advantage of large-number analyses averaged over monthly intervals can be seen in the smoothing effect that removes statistical variations and enhances first-order perturbations. The disadvantage lies in the eradication of any contributions due to weather.

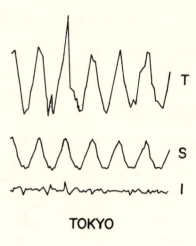

TOKYO

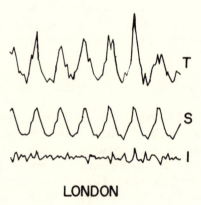

LONDON

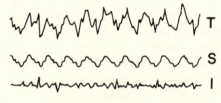

NEW YORK

Figure 10   Relative seasonal variations in heart failure deaths
over several years for different cities, indicating the
total death curve (T), the seasonal component (S), and
the residual due to nonseasonal variations in temperature (I). The peaks represent the winter months, while
the troughs indicate the summer periods.

Source: Modified from Sakamoto and Katayama (1971).

140

TABLE 1

Some Seasonal Variations in Various Human Measures

Measure	Maximums	Minimums
Behavioral		
Depression score (Minnesota Multiphasic Personality Inventory)	Fall (70)	Summer (60)
Psychiatric admissions (numbers)		
Texas	Summer	—*
Georgia	Summer	—
Illinois	May	—
Ohio	Spring (1,600)	Winter (1,400)
New York	June	—
Europe	Summer	—
Suicides (most states)	March through May	November through December
Blood		
Total protein	Winter	Summer
γ-globulin	>8.5 grams/100 cubic centimeter	<7.5 grams/100 cubic centimeter
	Summer	Winter
	>13 percent	<10 percent
Hemoglobin	Winter (December)	Summer (June)
Leukocytes	Winter (December)	Summer (August)
Eosinophils	March	July and August

(continued)

Table 1 (continued)

Measure	Maximums	Minimums
(Blood)		
Calcium	August (11 milligrams/100 milliliters)	February and March (8.5 milligrams/100 milliliters)
Phosphate	Summer	February
Iodine	July and August	December through April
Capillary Fragility	January through March	July through December
CO_2 capacity	December	June
Glands		
Thyroid	Winter (size & activity)	Summer
Adrenal gland (17 Keto-steroids)	Winter	Summer
Pituitary		
Thyroid stimulating hormone	Winter	Summer
Adrenal stimulating hormone	Winter	Summer
Gonad stimulating hormone	Spring	Winter
Gastroacidity	Winter	Summer
Disorders		
Mortality (Western Europe)	December through January	July
Bronchitis	Winter	Summer
Peptic ulcer	December through January	June

Adrenal ulcer	May and November	August and February
Glaucoma	Winter (November)	Summer
Goiter (simple	Winter	Summer
Goiter (Graves' disease)	May	Summer
Eczema	Spring	—
Herpes zoster	July	Winter
Infancy deaths	July (5,750)	February (4,250)

*Data not available.

Sources: Nonbehavioral measurements from Tromp (1963). Behavioral measurements from Gerbus and Dallara (1975) and Dalen (1975).

Figure 11 shows the relationship between month and average mortality from all causes in the United States over a 15-year period. Even if the variation in mortality were coupled with seasonal temperature, the maximum (January-December) and minimum (September-October) values would be only ± 10 percent of the mean annual values. No doubt, mortality is a multicontributory measure. Theoretically, one would expect only particular diseases to be coupled with seasonal temperatures.

Heart Failure and Stroke

Two general categories of mortality, heart failure and cerebrovascular accidents (stroke), have been correlated many times with ambient, seasonal (monthly) temperatures. The most common heart failure label has been myocardial infarct (MI). An infarct is an area of tissue in an organ that undergoes necrosis (cell death) following cessation of blood supply. Usually cessation in blood flow is associated with blocking (occlusion) or constriction (stenosis) of the supplying artery. A MI involves the formation of necrotic tissue within the heart muscle, thus reducing its total work potential.

A variety of contributing factors increase the probability of MI. Dietary, psychosocial, and genetic factors that influence the dilation and constriction of the coronary blood vessels supplying the heart contribute to this possibility. Some of these factors, such as the accumulation of cholesterol along the inner anterial walls, are progressive. Technically, any form of cardiac ischemia, that is, the temporary and local anemia due to obstruction of circulation, can contribute to the onset of MI. Many stimuli, including extreme, ambient temperature values, can contribute to ischemic reactions.

The most common term for a cerebrovascular accident is stroke or apoplexy. Stroke is associated with the formation of a thrombus that occludes a cerebral artery or with hemorrhage into the brain or spinal cord from a defective blood vessel. Frequently, this disorder is associated with arteriosclerosis or with the formation of an aneurysm. The latter is an arterial dilation due to pressure of the blood upon a weakened vascular tissue, resulting in the formation of a sac containing clotted blood. Sometimes the sac bursts, resulting in hemorrhage or release of the embol, and blocks a vital area.

Complications or mortality from stroke or heart failure are expected from both general theoretical understanding and empirical sampling of human physiology during temperature changes. Maintained elevations in ambient temperature severely tax the cardiovascular system due to the reflexive diverting of blood into the periphery.

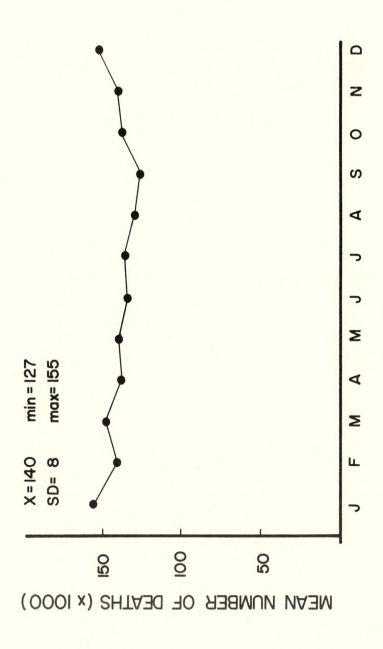

Figure 11   Mean absolute monthly variations in mortality from all causes in the United States between 1950 and 1965.

This process allows effective heat reduction of the blood by passage through the cooler skin. Unfortunately, this normally very adaptive response also elevates central blood pressure and produces marked vasoconstriction of coronary blood vessels and, possibly, some cerebral blood vessels.

On the other hand, during maintained exposures to cold ambient temperatures, more blood is forced within the body core to attenuate heat loss. The peripheral blood vessels constrict, thus reducing the surface area of the blood network exposed to the air, but the coronary blood vessels and probably some central cerebral blood vessels dilate, sometimes excessively. Blood pressure is maintained by a combination of altering heart rate, stroke volume, and vessel volume.

The Data

Seasonal variations in death from stroke and MI are strongly displayed in most North American, European and northern Asian countries. Correlation coefficients range between -0.8 and -0.95 for monthly mortality and monthly temperature averages over the year. Although this is impressive, one should remember that the correlation is dependent for the most part on the "ripple" riding upon a baseline of absolute death numbers. The absolute numbers of deaths from these disorders can display large ranges.

In New York City (1959-1962), death from heart disease ranged from 370 per 100,000 inhabitants in the warmest month (24°C) to 520 per 100,000 in the coldest month (-4°C). Although the correlation is -0.93 between heart disease and temperature (Sakamoto and Katayama, 1971), the temperature-related variation in death is ±20 percent around the mean. Similarly, London (1960-1964) shows minimal deaths of 300 per 100,000 (18°C) to maximums of 600 per 100,000 (3°C), while Tokyo shows minimums of 40 per 100,000 (28°C) and maximums of 60 per 100,000 (2°C). In all three cases, the range of variation is about from ±20 percent to ±30 percent of the mean death rate for that disease.

Stroke mortality shows similar ranges. Monthly mortality and monthly temperatures displayed correlations of -0.97, -0.92, and -0.91 in Tokyo, London, and New York City, respectively. Actual minimum-maximum values for these disorders (per 100,000) were 80-140, 90-230, and 80-100, respectively. Such ranges in absolute numbers of deaths strongly impress the importance of observing the raw data. The relevance of basic model assumptions, from the law of initial values to asymptote limits, should be apparent as well.

Katayama and Momiyama-Sakamoto (1970) have reported similar data for many countries. Not surprisingly, countries with smaller ranges in yearly temperature display steeper regression lines for temperature-death correlations than do countries with larger temperature ranges. Warmer countries demonstrate maximum death rates at the lower temperature extremes, even though these same temperatures in colder climates do not appreciably enhance death rates. The actual midpoint, at which 50 percent of mortality cases lie above and below the value, temperature varies from 12°C in southern Mediterranean countries to about 0°C in the northern European countries.

The striking and persistent negative correlation between monthly mortality and monthly or seasonal temperature is complicated by a variety of specific factors. Although one could become enthusiastic about discovering the biological bases of temperature-associated morbidity due to high correlation coefficients, the excitement is tempered by the complex geometry of internally checking the data. A variety of artifacts and confounding factors, especially age and type of disease, enter into the temperature effect.

In an illustrative study, Bull (1973) correlated monthly temperatures with monthly mortality rates in England over a three-year period. Unlike many other studies, Bull partitioned the data into smaller units in order to determine the actual source of the variance in the mortality associated with types of temperature change. Both maximum and minimum daily and monthly temperatures were correlated with the incidence of MIs, cerebrovascular accidents (such as strokes), venous diseases, and total respiratory diseases.

Such partitioning of the data into second- and third-order levels of analysis frequently displays a perplexing and complicating geometry. However, these analyses and complex geometries of numerical distribution allow the specific testing and isolation of mechanisms. Without sophisticated internal analysis of the data, confounding factors are never exposed and old myths continue.

Bull found the greatest correlations to be between temperature and myocardial infarct (-0.90). Comparable correlations were found between temperature and cerebrovascular accidents (-0.88). Deaths associated with all types of respiratory diseases displayed coefficients of about -0.75, while those associated with venous diseases showed coefficients of about 0.65. In other words, monthly temperature fluctuations accounted for 80 percent of the variability in MI deaths and cerebrovascular accidents, but only 56 percent and 42 percent of the variability associated with respiratory and venous diseases, respectively.

Closer inspection of the data indicated that the contribution of monthly temperature to MI was only significant for males over 55 years of age; females over this age showed a marginally significant relationship, while no significant correlation existed for either males or females under the age of 55 years. On the other hand, in the case of cerebrovascular accidents, only females demonstrated significant temperature dependence. Again, the significant effect was for subjects over 55 years of age.

Further analyses of the data indicated that minimum night temperatures demonstrated the highest correlation coefficients with the various mortality measurements. However, these differences were not statistically significant from minimum daytime temperatures when controlled for interdependence. Since meteorological phenomena, such as a stationary cold air mass, display half-lives in the order of two days, very low night temperatures are also associated with very low day temperatures.

Partial correlations indicated the interdependence of both MI in males and stroke in females with both respiratory disease and temperature. Bull found that when the variability associated with respiratory diseases was held constant, the relationship between MI and temperature was -0.67 (45 percent) for males and -0.79 (62 percent) for females. Similarly, the correlations for cerebrovascular accidents was -0.60 and -0.67 for males and females, respectively. To demonstrate that the effect was primarily temperature-dependent, Bull held the variability associated with temperature constant. As might be expected, the correlation between MI and respiratory diseases lost statistical significance (-0.26).

The problem of multiple and fractionated contribution to alleged single variables of effect (myocardial infarction) from a single weather stimulus (temperature) will be encountered again and again. In the case of Bull's data, the relationship between MI and maximum temperature was -0.80, while the relationship between pneumonia and maximum temperature was -0.80. The relationship between MI and pneumonia was +0.53 (or 25 percent of the variability in MIs could be accounted for by pneumonia cases and vice versa). Held constant, partial correlations with pneumonia indicated a correlation of -0.70 between temperature and MI, while the relationship between MI and pneumonia with maximum temperature held constant was -0.26 (not significant).

A preliminary interpretation of the data indicates that whereas a large part of the variability in MI cases can be accounted for according to the properties of temperature numbers, another portion of the pneumonia cases can be accounted for by temperature. However, the portions of both diseases related to temperature do not necessarily share the same properties of variability (presumably origin), otherwise they would be highly intercorrelated as well. Confusion is only an initial symptom of the passage from first-order inspection to detailed analysis of data; it will pass.

Other complications in monthly temperature contributions to human disorders are typified by the interaction of climate and confused diagnostic categories. Whereas acute MI cases increase within the colder months in most European and North American countries, persistent warm climates may show peak MIs during the hottest periods of the year. Heyer, Teng, and Barris (1953) indicated that the peak month for MIs in the cities of New York, Chicago, Philadelphia, and Boston were during November and December—an expected result on the basis of world data—but June for Dallas. In fact, the MI incidence in June was about 50 percent greater than for December (150 and 109). These data were not subjected to detailed analyses against the numerical properties of temperature measurements.

Time Lagging

The robust nature of the relationships between stroke mortality, for example, and monthly temperature averages can be seen by correlating the mortality measure with the temperatures for previous months. With lags of one month, correlations between stroke, death, and temperature drop from -0.90 for Tokyo, London, and New York to about -0.80 for these cities. With lags of two or three months, the coefficients drop from -0.50 to -0.15, respectively. Consequently, one cannot say that the stroke mortality correlation to temperature is purely confined to monthly factors.

The seasonal periodicity reflected in mean temperatures is especially important when weather effects are considered. Boyd's (1960) data for a variety of ailments in London between 1947 and 1954 demonstrate that the correlations between temperature and disease continue over weeks. Whereas the correlation between death from bronchitis and heart disease for men 45 years and older was greatest when analyses were completed with temperature the week before death (-0.72 and -0.74, respectively), correlation coefficients remained at levels of -0.50 or greater with temperature lags of up to six weeks before death.

Confounding Factors in Seasonal Temperature Effects

At first inspection, mortality data (especially MI and stroke tables) appear to display intricate dependence upon the monthly ambient temperature in most countries. The relationship is expected theoretically and has common sense appeal. One "knows" that temperature extremes are associated with death when the human organism is exposed directly to ambient conditions. However, these data are plagued by a number of confounding factors.

First, winter temperatures so frequently correlated with death are also associated with winter behaviors. During winter months, people of most cultures spend more time inside. Statistically, more living time is spent within high density human areas. Within these areas, the person is inundated with exhalant of hundreds of different people as well as with the consequences of artificial heating: drier air, more carbon dioxide, and less fresh air. Whereas air in the countryside may show only one bacterium per cc, up to one million bacteria per cc can be found in the air of department stores during mid-December.

One should expect, considering the nature of winter living space, a significant communicative (primarily immunological) contribution to the variability of heart diseases linked to temperature. This pattern is demonstrated somewhat in Bull's (1973) data. Respiratory infections, from the common cold to pneumonia, limit the amount of oxygen available to the blood supply and severely tax the heart muscle and vasculature. During periods of reduced oxygen uptake, the cardiovascular system can display sympatheticlike activity.

Under these conditions, the coronary vessels dilate and the cerebral vessels are stimulated, thus increasing the probability of stroke. Although lowered ambient temperature does produce similar changes, one must realize that relative hypoxia associated with respiratory inefficiency can produce similar responses. At present, we do not know which portions of the variability in winter mortality are a consequence of actual temperature, respiratory complications, or general immunological complications simply due to more human exposure during the constrained living environment of the winter season.

Second, winter weather is associated with more than just colder temperatures. Increased snowfall and general "nuisance" weather, such as sleet, cold rain, and muddy driveways, frequently require more attention and work. December and January in many countries are associated with the coldest temperatures and heaviest snowfalls. The consequences of mechanical strain from shoveling snow, pushing cars, and general ambulation would contribute heavily to heart overloading.

Third, winter weather restricts all forms of activity within the population (over 55 years of age) most reported to be affected adversely by low monthly temperature averages. During winter seasons, older populations are forced to repeat the same behaviors more frequently and to be exposed to the same stimuli. Whereas summer months contain a wealth of small but different stimuli—from colors of flowers to the brief sighting of a bird to the different combinations of people passing the observer's window—winter months reduce this variability to a monotonous minimum.

This singular psychological factor has a powerful impact upon the cardiovascular system, more than the average person might notice or admit. Experimental data indicate that repeated presentation of the same or similar stimuli and consequent boredom or depression can influence the cardiovascular system as extensively as many drugs. Within the vicious circle of winter boredom, repeated anxieties about personal death and ruminating thoughts of self-worth can induce protracted periods of autonomic stimulation to the heart and vasculature.

At present, no studies have grappled with this fundamental problem. Since one is dealing with populations of people over 55 years of age, how does one separate the cardiovascular consequence of temperature as a variable from the vascular consequences of winter-related behaviors? Indeed, the old members of most cultures are the ones most affected by winter temperatures; these people are also the ones most influenced by the psychological constraints of winter.

PROTRACTED PERIODS OF EXTREME TEMPERATURES

A more precise test of the effects of temperature upon human beings can be found during periods of prolonged extremes: heat or cold waves. These extremes usually occur for periods much longer than durations traditionally associated with weather changes. Usually these extremes are superimposed upon the peak or trough of the seasonal crest. Heat waves ride on top of the seasonal peak in temperature (July and August) while cold waves are extreme statistical excursions in temperatures below the usual mean (December and January).

From a systems model, one would expect heat or cold waves to evoke significant increases in mortality and temperature-correlated behaviors. Since the heat or cold waves are superimposed upon extremes in temperature values, one would further expect a nonlinear increase in the number of disorders as thousands of defective or weak homeostatic systems break under the extra strain. When pushed to the normal limits of temperature stimulation, just a few degrees more would be expected to fail the human system.

Heat waves and cold waves are excellent tests of temperature effects since they do not occur every year. Consequently, one can compare mortalities during years with normal seasonal variations and years with extreme temperature values. In addition, since heat

waves normally have life spans from about one to two weeks, the
specific influence of these extremes can be checked by measuring
the onset and offset time for the human measures. Presumably,
alterations in human behavior should be precisely linked in time to
the temperature extremes, rather than diffusely related to temperature values of the previous and consequent months.

The Heat Wave

During heat waves, temperatures can exceed 100°F (38°C)
for more than a week (at the crest) and surpass 90°F (32°C) for
more than 20 consecutive days. In North America, heat waves usually
occur during the month of July and, less often, during June and
August. Occasional miniheat waves are associated with the Indian
summer of September and October.

Heat waves have been recorded officially for at least 200 years
in the United States and appear on a semiperiodic basis. Some
authors have argued that they are coupled partially with solar activity.
In this century, notable heat waves occurred during 1901, 1911, 1916,
1934, 1936, 1952, 1953, 1954, 1955, and 1966. Usually these conditions are associated with extraordinary stationary air masses over
large regions.

Significant regional differences can exist for death measures
associated with heat waves. For example, St. Louis, Missouri has
been inundated with heat waves regularly. In 1854, the death toll
for one month (July) in St. Louis was comparable to that of the entire
year. Figure 12 compares the monthly death figures (from all causes)
for the United States as a whole and those for St. Louis, Missouri.
The discrepancy during July indicates the importance of considering
both regional and yearly differences in data analyses.

Physical Patterns of Heat Waves and Mortality

Since heat waves display precise intensities (temperature values)
and durations (time of occurrence), quantitative indexes can be derived for detailed comparisons with mortality numbers. General
inspection of the data indicate that heat waves of different magnitudes
exist along a statistical continuum ranging from miniheat ripples,
such as Indian summers, to epidemic waves. Their consequences
appear to be quantitatively, rather than qualitatively, different.

Most authors affirm that the average air temperature and the
number of successive hot days are the best predictors of heat-
associated death. As the mean air temperature and duration of

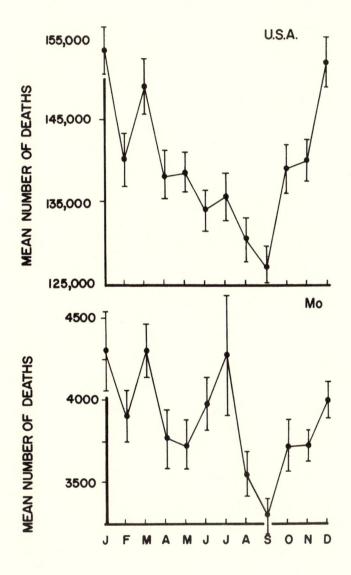

Figure 12   Mean monthly death numbers from all causes for
Missouri and for the entire United States.

population exposure increases, the number of deaths increases as well. It is not known, totally, whether the increase is always linear, curvilinear, or incremental ("quantal jumps"). Such a discrimination is important when considering mechanism.

Commensurate with a systems model, the greatest number of deaths predominate during the first heat wave of the season, even though later heat waves are more extreme. Peaks in death indexes usually occur on the day after the peak temperature of the heat wave. Usually the increase is a precipitous jump (sudden failure) and has been described as epidemic.

Types of Death Categories Associated with
Excessive Heat

In general, all types of deaths appear to increase with heat waves. Although the specific category of death by heat stroke and insulation has been introduced to the mortality tables since 1952, Ellis (1972) argues that all categories of death more adequately reflect the consequence of heat than those attributed to heat alone.

The majority of excess deaths associated with heat waves are those of people who already displayed histories of congestive heart failure, cardiovascular-renal diseases (hypertension), or cerebral vascular disease (such as cerebral arteriosclerosis). The acute cause of death is usually heart failure, electrolytic imbalance, dehydration or cerebral hemorrhage. The latter condition has predominated in past heat waves.

Deaths from respiratory ailments do not demonstrate excessive increases during heat waves according to most analysts. Other mortality categories, such as cardiac ischemia or accidents, increase occasionally. Interestingly, death from homicide does not show a systematic increase with heat. Some years are associated with increased homicides following heat waves.

Other diagnostic categories appear to contribute to heat stroke/ death risk. The incidence of death due to diabetes has been reported several times. During the heat wave of July 1966 in New York City, the number of deaths associated with diabetes increased 81 percent over expectancy. Other well-accepted risk factors that predispose a person to heat stroke are fatigue, lack of physical conditioning, failure of acclimatization (such as in congestive heart disease), and obesity.

Extremes of core temperature can generate the vicious circle effect. As the core temperature rises to above 37°C, the basal metabolic rate (BMR) increases 10 percent for every 1°C due to the intrinsic effects of heat on enzymatic reactions. Unfortunately, there is a limit to the amount of heat loss through sweating, a condition aggravated markedly by dehydration. At some point, the excessive heat directly and adversely affects the neurons of the hypothalamus, reducing sweating even more. Suddenly, the vicious cycle precipitates.

Above about 106°F core temperature, the basic protein structure composing many cells deteriorate, most of the time irreversibly. Although many organs still contain the capacity to replace or reproduce cells, areas such as the brain do not. Permanent damage ensues. The loss of cell wall characteristics is especially important in the vasculature. Breakdown of the cells that compose the blood-brain barrier result in a myriad of minihemorrhages that can lead to a sudden deluge.

The Magnitude of the Effect

Few researchers have seriously studied the magnitude of effect associated with heat waves. Considering the complexity of analyses, the problems of gross diagnostic lumpings, the great range in local temperatures, and the extraordinary differences in baseline of heat-related diseases, one is not surprised. Depending upon the locality for example, death from heart disease ranges from less than 60 per 100,000 to well over 300 per 100,000. Gross lumping of heat wave values from areas with both great and small incidences of these disease would certainly mask actual patterns.

Few recent studies have demonstrated the extreme death rates (mortality in one month equal to that of year) reported for heat waves in the previous century. During the heat wave of 1901, 4,000 of 518,000 total deaths were associated with excess heat; during a similar period in 1902, only 290 of 508,000 total deaths were so associated. Although the heat wave-related death rate was more than 10 times that of a similar period the following year, its absolute contribution to the total mortality of the year was less than 1 percent.

According to Ellis' (1972) data for the heat waves of 1952-1955 and 1966 (five years), stroke demonstrated significant increases during the heat wave months of July. Whereas stroke numbers peaked for these years in January and December (16,000 = mean) as expected, the trough was in August and September (mean = 13,000). July of these years had 15,000 stroke deaths, about 1,000 (~10 percent) more than expected from simple extrapolation of the curve.

Within the five-year cluster, significant heterogeneity existed. Stroke deaths actually dropped from 13,981 during June of 1953 to 13,032 during July of that year. On the other hand, stroke deaths increased from 13,505 to 15,372 for June and July of 1955. For

1966, the stroke rate increased from 16,357 to 17,990 for June and July. In both cases, the increases were about 10 percent over the expected baseline.

The differences can be accounted for somewhat by the absolute differences in the actual high temperatures recorded during these years. Higher death rates were associated with higher temperatures and longer durations of the heat waves. However, not all the variance can be accounted for within these terms.

COLD AND WARM FRONTS

The temporal existence of warm and cold fronts, that is, the interface between two air masses containing the maximum temperature gradients, ranges between two hours and two days over the measurement area. Air masses, on the other hand, may exist over a measurement area for days and sometimes weeks on rare occasions. In both cases, the terms warm and cold are relative to the baseline upon which they are imposed. The absolute temperature of a cold front during the winter is much different than the absolute temperature of a cold front during the summer.

Assuming a linear relationship between biological effect and temperature change, one would not expect very significant changes in human populations from temperature changes associated with weather. Variation of monthly average temperature over the year in most areas analyzed in temperate regions of the northern hemisphere is about 40°C. Since the seasonal variation of mortality measures is about ±10 percent of the annual mean, one wonders whether the from 5°C to 10°C temperature changes associated with daily weather (not day/night changes) influence similar disorders.

Theoretical Consequences

Temperature fluctuations associated with weather changes are important in the context of two variables: time and baseline. Indeed, the annual average range in temperature may be 40°C in temperate regions, but this occurs over a six-month period. On the other hand, the 10°C change in temperature associated with both fronts and air masses occurs within three days. Whereas the system could adjust to slowly altering temperatures, the sudden change in temperature, even so small in magnitude, might tax some systems more extensively than even large-magnitude effects over the seasons.

The time component is the key to understanding the effects of temperature upon human beings. Habituation is a property of living systems. Given sufficient duration, even simple single-cell systems can adjust to perturbations. However, scientists are only beginning to appreciate the consequences of small perturbations with short durations.

Baseline considerations are also important. If the boundary for stability within the organism changes with the seasonal baseline in temperature, temperature fluctuations would be expected to be minimal stimuli (Figure 13A). If the boundary for stability did not change with the seasons, small perturbations in temperature during the periods closest to the boundaries of stability (winter and summer) would be expected to be the "straw that breaks the camel's back" (Figure 13B). Like other portions of the weather matrix, studies with temperature should be done within the different seasons.

Organismic Responses

Sudden decreases (a few hours) in ambient temperature (10° C) evoke increases in clotting speed, peripheral vasoconstriction, urination volume, 17-oxytosteroid excretion, capillary resistance, muscle glycogen, blood pH, blood pressure, blood sugar, and relative hypoxia. Sudden drops in temperature evoke decreases in blood volume, permeability of tissue, and peripheral capillary space.

Elevations in temperature evoke opposite reactions in normal people. Peripheral vasodilation is associated with decreased capillary resistance, urination volume and muscle glycogen, but with increased blood volume, relative hyperoxia, and tissue permeability. The other measures show similar changes. During heat extremes, however, adrenal corticosteroids and thyroid activity may increase as well. Due to the intrinsic properties of heat upon chemical reactions, extreme heat may elevate BMR.

During cooling, pH of urine increases, sometimes from 5.0 to 6.5 over a 24-hour period (Tromp & Bouma, 1966). The excretion of hexosamine (one of the mucopolysaccharide components) decreases, either from retention within the body or less synthesis of the precursor molecule, with increased cooling. During strong cooling associated with a cold air mass, hexosamine excretion may drop to 20 microgram (μg)/100 milliliters urine, while during warm air periods, it may rise to 70 μg/100 milliliters in normal patients.

Temperature, the Common Cold, and Influenza

A number of studies have indicated significant relationships between ambient daily temperatures and the common cold. Several

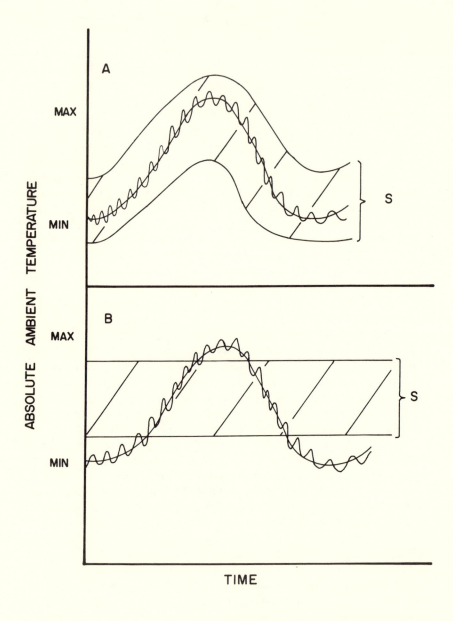

Figure 13   Two conceptual conditions in which the stability limits for a human system are either flexible to temperature change (A) or inflexible (B). Whereas A would still allow system stability, condition B would evoke periods of instability during the extremes of the temperature values.

158

experimenters have indicated that the highest correlation between these two factors is the temperature two to three days before the onset of the cold symptoms. For the most part, correlation coefficients are negative; that is, as temperature drops, cold ailments increase.

The common cold is not really a symptom but a syndrome or combinations of symptoms. Severity of the cold, as defined by physiological disruptions (hypo- or hyperthermia, constipation), behavioral alterations (difficulty in concentration), and duration of symptoms, ranges from one day to over three weeks. The average duration appears to be about one week. Enhanced severity of the responses and recruitment of other symptoms (nausea, and so on) usually blend into flu or influenza-type syndromes.

The immense variety of sources that can induce common cold/light flu symptoms (sometimes the differences are indistinguishable) has been used to explain the apparently inconsistent relationships between these symptoms and weather. As many as 100 different viral conditions appear to be related to common cold symptoms, and these can be influenced by the baseline resistance of the person (which decreases after the winter months), the condition of the air breathed (low humidity, toxicants, pollutants), and any factor that can momentarily reduce the normal homeostasis and defenses within body tissue. Sinuses, nasal-pharynx, larynx, and lungs are statistically likely areas of impact because of their direct interfaces with heterostatic stimuli from the air.

Cassell, Lebowitz, Mountain, Lee, Thompson, Wolter, and McCarroll (1969) monitored 1,747 people over a 45-week period and compared their daily ratings of cold symptoms with daily weather measures, including pollution concentrations. The most significant correlations existed between temperature and the numbers of colds (-0.61), coughs (-0.50), and sore throats (-0.45). Headache and eye symptoms were not significantly correlated with these measures. The other meteorological measures of precipitation, wind velocity, relative humidity, and barometric pressure as well as the pollution indexes of carbon monoxide, total hydrocarbons, and sulfur dioxide displayed correlations of less than +0.30 (less than 10 percent of the variance) for any of the headache, eye, colds, cough, and sore throat scores.

However, single variable correlation may not display all the potential of the weather matrix. For this effect to be unmasked, more sophisticated analysis, such as multiple regression, must be used to look at the amount of variance in the behavioral measure that can be explained by combinations of the different weather variables. Using the ten variables of the last paragraph, Lebowitz (1973b) found that the common cold syndrome correlated with the following variables as a group: particulate matter (+0.30), temperature (-.81), relative humidity (+0.20), and barometric pressure (-0.18). The multiple correlation coefficient was 0.93 (86 percent of the variance in common cold explained). Note, however, that temperature was still the major predictor variable.

Lebowitz's data also demonstrate one other important point about the temporal variability of weather contribution to common cold incidence. When the nine seasons between winter 1962/1963 and winter 1965/1965 were analyzed separately for the contributions of particulate matter, carbon monoxide, sulfur dioxide, wind velocity, solar radiation, relative humidity, temperature, and barometric pressure, the multiple correlation coefficients ranged from 0.34 to 0.93. In other words, the range in amount of variability in common cold incidence associated with these weather variables ranged from 12 percent to 86 percent.

Between the spring of 1963 and the spring of 1964 (five seasons), correlation coefficients between temperature and the common cold ranged from 0.24 to 0.81. Correlations for barometric pressure, on the other hand, varied from 0.26 to -0.57 for the six seasons in which the coefficients were significant. None of the weather or pollution measures showed significant correlations with common cold incidence for all seasons measured.

The possibility that weather changes can trigger epidemics has been implied, theoretically at least, by several authors of the past. The two basic models used have involved a type of cascading concept. It presumes that with optimal immunological resistance (or its lack) in the community, the probability of a sudden epidemic requires some large area stimulus in addition to simple statistical interaction of human beings in crowded places.

Donle (1975) noted that over a 59-year period, the development of influenza during the year (primarily January through March) demonstrated an impulselike curve typified by sudden increases, such as 100 to 8,000 cases within about two weeks. He was further impressed by the first-order observation that onsets of influenza epidemics in different parts of Germany, Norway, and Switzerland were often synchronous. Usually, these increases occurred about two weeks after sudden influxes of cold air over western-northern Europe.

Donle argues that the two-week lagging procedure is essential to appreciate the intense relationship between the onset of the epidemics and weather. One of the most common types of weather patterns associated with the onset of influenza patterns was a cold air mass separated by a warm air mass and then a colder air mass. By March, at least in that part of the world, when warmer air masses become more predominate, influenza incidence is markedly reduced.

Mortality

Whereas the relationship between monthly temperature averages and monthly mortality is well known, the correlation between daily mortalities and daily temperature measures is subject to much greater variability. With such fine temporal discrimination (one day versus one month), one may have lost the optimal "window of analysis" by constricting the aperture too much.

Daily analyses reduce the probability of determining any effect from temperature since mortality often shows differential lagging. Whereas sudden drops in temperature may evoke increased mortality the following day, other temporal patterns of temperature change may evoke mortality increases from one to four days later. Even if all major temperature relationships were to occur within four days, daily analyses could reduce the magnitude of the effect.

With this limit, Lebowitz (1973b) has found significant correlations between daily temperature and mortalities in New York City. The winter of 1962/1963 showed a surprisingly significant -0.74 correlation between daily mortalities and temperature. Other seasons did not show this effect. Other experimenters have been less successful with these analyses.

Boyd (1960), on the other hand, who used weekly averages for deaths and temperature (for London), did report lag effects. According to his analysis, the correlation between death associated with respiratory infections (primarily pneumonia and bronchitis) was highest with the temperature two weeks before death. Whereas the correlation between death and temperature of the same week was -0.5 for people over 45, it rose to -0.75 when correlations were made between death rates and the temperature one and two weeks before. Further lags, up to -6 weeks, still showed maintained, elevated negative correlations (~0.50) between temperature and mortality.

Eclampsia

Eclampsia is a sudden convulsion, similar to an epileptic attack, that does not originate from obvious nervous pathology. It is a major toxemia of pregnancy, usually accompanied by high blood pressure and several nonspecific urinary changes. Although its cause is unknown, eclampsia appears to be more common in first births, multiple pregnancies, and in patients with anemia or malnutrition.

Neutra (1974) reviewed the literature between 1895 and 1973 and found 36 data studies. While 6 reported no correlation with

weather, 15 of the 36 studies noted that the disorder was most common during cooler or cold weather. In order to extend these studies,
he evaluated the relationship between weather and eclampsia among
156 patients in Colombia.

The numbers of eclampsic cases were twice as high on cool,
humid days than on days with average temperature or humidity.
Eclampsic rates changed from 8 per 1,000 hospital deliveries with
maximum temperatures of 25°C to 2 per 1,000 with maximum temperatures of 32°C. Similarly, at relative humidities greater than
80 percent, the rates were 10 per 1,000, but dropped to less than
2 per 1,000 at less than 68 percent relative humidity.

Critical Duration of Exposure

People in the western culture have developed a technology to
maintain relatively controlled living temperatures. During the
winter, most people, especially those over 55 years of age, spend
at least 90 percent of the day within some stable temperature around
$21 \pm 2°C$. A multitude of different energy-forming technologies,
from merely adding extra clothes to turning up the thermostat, have
allowed this temperature constancy.

During the summer, temperature control within living quarters
has been more difficult. Not only is the removal of heat energy from
the environment limited by the necessity of a more complex technology, but the organism itself is limited to a single means of cooling
(vasodilation-evaporation). However, in the last 20 years, the
development of air conditioning and other cooling procedures has
produced a more constant temperature during extreme summer temperatures.

If the population spends most of its time in a controlled temperature environment, especially during the winter, but still displays
temperature-related response curves, what is the mode of operation?
Are the temperature-related effects due to some indirect change,
such as humidity alterations, within the household? Or is there a
critical period of exposure to outside temperatures?

Rephrased, what is the threshold duration of exposure to the
outside temperature that will evoke discernible changes in a person?
What is the minimum time of exposure to the outside temperature
to which the human being would respond with a maintained period of
instability? How long would this instability last, and how much would
it contribute to the further development of a disease condition?

If the inside temperature were 20°C and the outside were -10°C,
would only 15 minutes of exposure produce a response? Would similar biological changes occur if the person were exposed for two hours

to a cool draft that reduced the effective room temperature to 18°C?
Numerical answers to such questions are necessary data for a complete understanding of the relationship between seasonal, monthly,
or weekly temperatures and human behavior.

HUMIDITY: WATER VAPOR IN THE AIR

Water vapor is the most variable of the gases in the atmosphere,
ranging from almost zero to a maximum of about 4 percent by volume.
Such variability is important to human existence, since water vapor
contributes not only to cooling (evaporation) and heating (condensation), but also to the distribution and extent of precipitation over the
earth.

The amount of water vapor in the air is determined not only
by the local availability of water from bodies of water or from potential precipitation, but also from the temperature of the air. As the
temperature increases the amount of water vapor that can potentially
be stored within the air increases in a systematic manner.

The degree of increase is best depicted by actual data. Using
saturated steam as an example, the amount of water vapor contained
within 1 meter (m^3) at 0°C is only about 4.8 gram (g), while at 10°C
the amount of water vapor rises to 9.4 g. In more typical living
environments, such as 20°C, 25°C, or 30°C, the amount of water
vapor in saturated steam is 17.3 g, 23.0 g, or 30.4 g, respectively.

Measures of Humidity

Humidity can be measured according to absolute or relative
scales. Absolute humidity is the actual mass of water vapor per
volume of air. Relative humidity (RH) is the percentage of water
vapor present in the air in comparison with saturated conditions.
If saturated air at 20°C contains 17.3 g/m^3, air at 20°C containing
only 15.0 g/m^3 displays the property of 87 percent RH. Alternatively, 50 percent humidity at 20°C indicates that the air contains
8.7 g/m^3 (50 percent) of water vapor.

It should be evident that air with a RH of 70 percent at 10°C
does not contain the same amount of water vapor at 30°C. The
absolute difference in water vapor in the air is quite significant.
Such differences are important when considering biological functions
such as cooling and hydration.

A common measurement for humidity utilizes wet-bulb and
dry-bulb temperatures. The usual procedure to obtain these measurements involves two similar thermometers, one of which has a thick

piece of clean muslin tied around the bulb. The bulb of the muslin-wrapped thermometer is dipped in water and the two are whirled or slung through the air. After a minute or two of whirling, the two thermometers are read.

The wet-bulb thermometer will display a lower temperature than the dry-bulb thermometer. The cooling of the mercury in the wet-bulb thermometer is due to the evaporation of the moisture around it, a process that involves the absorption of heat energy. Since the amount of evaporation is proportional to the dryness of the air, the difference between the dry-bulb and wet-bulb temperatures can be used as a measure of the moisture or dryness of the air.

The drier the air, the more the evaporation around the bulb and the greater the depression of the wet-bulb thermometer. With an air temperature of 70°F, the wet-bulb is depressed 1°F, 6°F, or 16°F at RHs of 95 percent, 72 percent, and 33 percent, respectively. At an air temperature of 100°F, 95 percent, 72 percent, and 33 percent RHs are associated with wet-bulb depressions of 1°F, 8°F, and 23°F, respectively.

Relevance of Humidity

A parcel of cold air at 0°C and 50 percent RH contains about 2.4 g of water vapor/m^3. If the same parcel of air were warmed to 20°C, without adding any water vapor, the RH of the air would drop to 15 percent, since the potential saturation point of the air for water vapor would be 17.3 g/m^3. Even if the initial 0°C air were near RH of 100 percent, the RH would be only 28 percent if the air were heated to 20°C.

Certainly this basic physical/chemical phenomenon describes a typical wintertime problem in most family dwellings. Outside, the temperature of the air averages around 0°C with RHs around 50 percent. The air is moist for that temperature. However, by the time that same air has been heated to room temperature (20°C), the RH has fallen to a very dry level. Relative humidities of less than 20 percent are not uncommon in many houses during cold winter months.

Very low RHs facilitate the display of electrostatic discharge since charged particles can be stored in appropriate areas without dissipation through water droplets in the air. The mild shocks associated with walking across carpets, sliding across seats, or even taking off some types of clothing have more nuisance effect than significant biological impact. The direct influence of ions associated with these conditions will be considered separately.

The primary effects of very low RHs, especially during the winter, are upon the nasal-pharynx and, in some cases, the trachea. Regardless of the initial ambient humidity and temperature, a parcel of air has been raised to 30°C (86°F), close to saturation point (100 percent RH), by the time it reaches the pharynx. The consequent elevation in body temperature occurs within the trachea.

Strain can be placed upon these "hydrating" passages when very cold air is raised to body temperature. If the ambient air were 70 percent RH at 0°C, the RH would have dropped to 11 percent $(3.4 \text{ g}/30.4 \text{ g}/m^3)$ by the time it reached the 30°C of the pharynx. The deficit in water vapor to maintain near saturation must be accommodated by the tissues themselves, primarily the nasal passages and the upper respiratory tract.

Simple calculations indicate that the amount of water taken from the nasal tissues to compensate for very low RHs due to the heating of cold air can be significant. In the previous situation, the nasal passage tissues must supply 27 g of water/m^3 of inhaled air at 10 percent RH in order for the air to reach saturation. Since the average person can breathe about 10 m^3 of air per day, the body must lose 270 g (270 cc) of water per day in this region alone.

Breathing directly through the mouth burdens the throat, trachea, and upper respiratory tract even more, since the heat energy and water vapor required to elevate the ambient air must be taken from these regions. Although these areas are subject to recurrent aqueous supplies, excessive dehydration of these regions can induce transient heterostasis and can increase the likelihood of viral or microbial development if viruses or microbes are present.

At present, little research is available concerning the statistical likelihood of respiratory complications as a function of the amount of moisture required from the body per day. The role this source of water loss contributes to whole-body dehydration, especially in the wintertime, is not clear either. This factor would be important since it is possible that the rate of removal may not be easily detected by osmoreceptors in a manner associated with greater increments of water loss.

Considering the glaring discrepancies between RHs in the outside air and the RH of heated dwellings during winter months, one should immediately be cautious about the significance of correlations between outside RH values and disease during this period. Humidity research requires simultaneous measurement of RHs for the ambient parcel of air and for the RHs of the average dwelling. This source of variability can be quite significant, especially for behavioral disorders.

The potential nature of humidity as a confounding variable in seasonal mortality rates should be considered seriously in this context. One of the most clear differences between cardiovascular and stroke mortalities associated with summer temperatures and with winter temperatures is the complications from respiratory ailments during the winter. No doubt, temperature has significant effects as a singular variable. However, partial correlations indicated that a significant portion of these temperature-related deaths in the wintertime were associated with respiratory infections, especially bronchitis and pneumonia.

Excessive humidities in warm air can also produce difficulties. Air parcels with temperatures of 30°C and 80 percent RH have as much as 24 g water vapor per m^3. Whereas drier air parcels at the same temperature would contribute to water loss from the body surfaces and hence evaporative cooling, high RHs would dampen this process since the air is already close to saturation. Possibility of heat stress would increase.

Humidity and Respiratory Complications

From the preceding discussion, one would expect that nasal and upper respiratory conditions would be related conspicuously to extremes in RH, especially low RHs. The relationship between hay fever, varieties of the common cold, and, to some extent, bronchitis and pneumonia have been suspected for several decades (for example, Unger, 1953, 1950).

Correlational studies by George Green (1966) in several Canadian schools clearly indicate that winter absenteeism is related to RH. Relatively striking correlations were found between absenteeism in six Saskatoon schools ($r = -0.92$) and six Halifax schools ($r = -0.75$). The actual range of the RH was from 20 percent to 40 percent in the Saskatoon setting and from 20 percent to 30 percent in the Halifax environment.

The magnitude of the effect on absenteeism is small but impressive, considering the number of students involved. Absenteeism increased from about 3.5 percent of the population at RHs of 30 percent to about 5 percent at RHs less than 20 percent. Although small in magnitude, this would correspond to the difference of 350 versus 500 absentees in a population of 10,000 students.

Relatively inexpensive hydration techniques with humidifiers can alter disease frequencies in a statistically significant manner. In one study, absenteeism was 5.7 percent in the control group maintained at the usual 40 percent RH, while the group exposed to RHs of 50 percent (both with air temperatures of 72°F) was 3 percent. Neither group was told the nature of the experiment.

In another situation, different barracks were either artificially humidified to 40 percent RH or maintained at the unusual level of 20 percent ± 8 percent RH (inside air was 71°F ± 1°F). A close surveillance of the health records showed no real differences in the numbers of gastrointestinal, genitourinary, skin, laceration-contusion, or psychological ailments between the recruits housed in the humified or nonhumidified barracks. However, the incidence of respiratory infections of all types decreased from 1.33 to 1.14 per trainee.

The attenuation of respiratory incidence is relatively trivial from an absolute perspective. However, it should be realized that the subjects were young and, in general, resistant to complicating respiratory infections. One wonders how much humidification would influence the frequency of respiratory infections in people over 55 years of age or in populations who have more than five "colds" per year.

The actual mechanism for humidity contributions to respiratory diseases can be implied from the water requirement of select tissue or from their changes in fluid viscosity. Richards and Marriott (1974) noted that the viscosity of bronchial mucus increased with lowered humidities. Such increase in viscosity reduces the efficiency of oxygenation of surrounding tissues as well as their removal of metabolic materials. In some situations, excessive mucous viscosity could prevent the phagocytic process from attacking materials embedded in the mucus.

In light of these data, one would expect a significant effect of RH on the infectivity of airborne viruses. One fine example of this effect was reported by Lester (1948), who exposed mice to influenza "A" virus delivered through a spray for 15 minutes. During the experiment, the mice were exposed to various RHs ranging from 10 percent to 90 percent (22°C). After 14 days, the surviving mice were killed and examined for pulmonary lesions.

Mortality figures indicated a clear curvilinear relationship between RH and mortality. Almost 100 percent mortality occurred in mice that had been exposed to RHs below 35 percent or above 80 percent. Mice exposed to RHs between 45 percent and 60 percent displayed about 20 percent mortality. A similar pattern existed for pulmonary lesions, except that the lowest incidence (50 percent of the population) of lesions occurred within 40 percent to 55 percent. All animals below 40 percent or above 60 percent RH showed pulmonary lesions.

These data demonstrate, for this virus at least, that the difference between 45 percent RH and 30 percent RH is the difference between life and death in infected mice strains. With respect to infection and pulmonary complications, a less extreme consequence, the difference between 50 percent RH and either 60 percent or 40 percent

RH, only a 10 percent range, was the difference between a probability of 0.4 and of 1.0 for developing respiratory infections.

Certainly, people are not routinely exposed to such potent viral conditions. The general impressiveness of the study must be tempered by the different survival curves of different viruses as a function of RH. Whereas rous sarcoma displays less than a 10 percent survival rate at RHs around 40 percent $\pm$ 5 percent and more than 90 percent survival at RHs less than 20 percent or more than 60 percent, staphylococcus displays between 30 percent and 40 percent survival in RHs of from 10 percent to 60 percent and increases between RHs of 60 percent and 80 percent (70 percent survival). Pidgeon pox maintains at least 80 percent survival regardless of RH.

Comfort

Relationships between the sensation of comfort, relative humidity, and temperature are well known. The humidity of the air affects a person's sensation of warmth. At a given temperature, increasing the RH increases reports of warmth. The effect is more marked at higher (greater than 25°C) temperatures, as the process of evaporative cooling is inhibited.

Experimental literature concerning subjective comfort, RH, and temperature indicates the same general trends. Often the absence of apparent detection of different humidities by people reflects the limits of the type of measurements used rather than the organism's potential response. Koch (1963) reported an experiment in which sedentary subjects were exposed to a range of RHs between 15 percent and 95 percent for three hours at a constant air temperature of 25°C. Although the subjects were not able to detect low humidities with any reliability, half the subjects ranked normal RH between values of 30 percent and 50 percent and high RH between 76 percent and 95 percent.

In a more recent study, McIntyre and Griffiths (1975) exposed subjects to one of three humidities, 20 percent, 50 percent, or 75 percent, at one of two room temperatures, 23°C or 28°C, for six hours. During and after exposures, the subjects recorded their impressions upon a number of different affective evaluation scales, each ranging between 1 and 7. Personality tests to measure introversion/extraversion were also given.

The following scales were used: open/oppressive, no discomfort/discomfort considerable, dry/moist, hot/cold, invigorating/dozy, steady/changing, dry/moist eyes, dry/moist skin, clear/congested nose, and little/no sweating. Subjects ranked their experiences several times during the setting. Each ranking series was placed into a box so that the subjects could not refer to previous scores.

These experimenters found that at 23° C, subjects could distinguish both high and low RHs. Both 20 percent and 75 percent conditions were considered more oppressive and uncomfortable than 50 percent. At 28° C, the lowest RH (20 percent) was preferred, and higher RHs, including 50 percent, were more uncomfortable and more moist. Personality scores did not appear to significantly influence the ratings.

Actual scores, although significant, were well within the low-level range expected for these gross descriptors of human experience. At 23° C, the low, medium, and high relative humidity environments were associated with average ranks of 3.3, 2.6, and 3.5 on the open (1) to oppressive (7) scale, respectively. The dry (1)/wet (7) rating averages were 3.2, 4.4, and 5.3, respectively. As might be expected with sedentary people, all aversive polarities of the scales increased as the length of time spent in the experimental situation increased.

A more general application of the relationship between temperature, relative humidity, and comfort has been quantified by meteorologists. Thom has derived a discomfort index that can be calculated with the following formula: $DI = 0.4 (t_d + t_w) + 15$, where t_d = dry-bulb and t_w = wet-bulb temperatures. People report discomfort as the DI rises above 70. Almost everyone will report discomfort when the index reaches 79. As the index passes 80, discomfort becomes acute, even serious (Thom, 1959).

SOME SPECIAL CONDITIONS

There are several everyday phenomena in the weather matrix that frequently escape evaluation. These potential stimuli are either too diffuse to evaluate or do not have a routine instrument for measurement. Whereas most of the phenomena would contribute clearly to the energy balance within the body, others would influence (in as yet unspecified patterns) a person's perceptual responses.

Rain

The amount of precipitation can be recorded easily and is a routine part of weather station data. However, the variable but significant contribution of rain to heat energy loss from the human body is more difficult to evaluate. On a common sense basis, most people would agree that "being soaked to the skin after being caught in the rain" may cause the onset of a cold. Whether the correlate is real or superstitious has not been verified by data.

Average dry clothing traps a small amount of air between the skin and the cloth material. When warmed, this air and the clothes act as insulators to attenuate heat loss due to radiation from the skin and from the conduction across the small volume of air. A usual suit of clothes attenuates heat loss to about half that value emitted from the same body when naked.

Damp or soaked clothing no longer insulates heat effectively. The small volume of air is replaced by water, which, because of its high conductivity for heat, increases the rate of heat transmission by about a factor of 5 to 20. Depending upon the local temperature and wind velocity, heat loss can become serious. This problem can be fatal, within minutes, in the arctic.

Data must still be collected to determine whether short periods of environmental exposure in wet clothes contribute to illness. What is the critical threshold of exposure? Is it related to the simple discrepancy between the body temperature and the environmental temperature? Or is the presence or absence of the appropriate pathogen the critical feature?

Cloudy Days

The percent of cloud cover can range theoretically from 0 percent (no visible cloud present) to 100 percent (complete cloud cover). However, if the continuum were divided into 10 percent increments, the distribution of each of the 10 increments as a function of time would not be equal. With respect to total time, the from 0 percent to 20 percent and 90 percent and above categories would predominate.

Such fine discriminations may be more relevant to behavioral studies than to heat physiology. From this perspective, the importance of cloud cover is linked almost totally to incident energy variance. Cloud cover influences the daily sums of global radiation very significantly. Whereas cloud cover as high as 50 percent influences radiation in a marginal manner, more dense covers retard radiation in a nonlinear manner. Complete cloud cover can reduce incident radiation by 50 percent.

Since the incident radiation can change over the year (and during the day), the contribution of cloudiness can be quite significant in an absolute sense. For example, in Missouri, the daily sums of global radiation range from 200 (gcal)/cm^2 in January to 600 calories cal/cm^2 in July (Tromp, 1963). Typical daily variations in radiation range from ~0 (during the night) to 0.8 gcal/cm^2/minute for January and from ~0 to about 1.5 gcal/cm^2/minute for July.

Depending upon the conditions, certain cloud covers can enhance the amount of heat and radiative energy between the earth and the cloud surface. Since the human being readily absorbs infrared radiation, regardless of skin color, some cloudy days can be associated with a marked enhancement of heat energy. In conjunction with high humidities and temperatures, heat loss can become even more difficult despite cloudy skies.

Brightness

Brightness refers to the amount of visible light striking the observer's environment. Although visible light intensity contributes little to heat energy loss or gain, both are influenced by similar conditions. Since variability in brightness is frequently linked by time to variability in heat availability, conditioning and confounding factors are real problems.

The effects of brightness alone, as a phenomenon, upon biometeorological events has received little attention. Whereas the intensity of visible light and its duration have been repeatedly associated with seasonal behaviors, such as reproduction and migration, the consequences of light brightness associated with weather have not been considered.

In principle, variations in light intensity associated with weather could directly stimulate the same mechanisms coupled to seasonal variations: the autonomic nervous system. Intense bright light evokes constriction of the pupil, which in turn evokes a number of different autonomic responses. The general consequence of sympathetic dominance is increased overall activity in most response systems—from ambulation to feelings of well-being. More intense conditions, when maintained, can contribute to headaches, nausea, and general malaise.

Low light intensity is associated more with parasympathetic dominance. Within a low brightness environment, the pupils dilate. In general, depending upon both learning history and neurophysiology, this condition is associated with relaxation, tiredness, sleepiness, and the initial stages of sexual arousal. Nibbling behaviors and general depression are more extreme changes.

Within the night domicile, the consequences of different light intensities are clear but, again, confounded by basic neurophysiology and learning. Some students report a required threshold of light in order to maintain studying behavior. At intensities of 2,200 lux, typical of a well-lit room (overhead fluorescent lamps and extra reading lamps), they report maintained interest. Below 700 lux (one dull reading lamp), they report sleepiness and general loss of concentration within about 30 minutes.

The change of incident light within the household on successive clear or cloudy days can be quite significant. For example, if one stands 3 meters from a 6 meter by 6 meter picture window when the sun is directly overhead, the light intensity in the room can range from 1,200 lux on sunny days to 400 lux on cloudy days.

By merely walking outside on a bright day, one can change the incident light intensity upon the body surface from 300 lux to 30,000 lux, depending upon the objects immediately in the visual field. In the same outside environment, on a cloudy day, the light intensity may be only 2,000 lux.

Since both learning and reflexive mechanisms are involved, one would expect simple habituation curves to describe the relative effectiveness of light intensity changes upon the autonomic nervous system. A bright day following three days of dim, cloudy conditions would evoke a much greater response than the third of three successive bright days.

SOME GENERAL BEHAVIOR NEUROBIOLOGICAL EFFECTS OF TEMPERATURE

Despite the vast literature on temperature effects, a general theory for the mechanism by which weather-related temperature changes influence the human organism is still speculative. The problem lies in the relatively low-level nature of the temperature range associated with classic weather changes. Whereas acute, exaggerated exposures to temperature extremes may evoke clear responses, the complex nature of temperature changes becomes a real statistical problem.

Human report adds little to the solution since ambient weather-related temperature changes may mediate changes in systems not correlated with consciousness nor specific word labels. At effective skin temperatures near 33°C, cold and warm receptors are continuously active, but no conscious temperature sensation is reported. The latter begins only when a relatively high number of thermal impulses per unit time reaches the central nervous system (Hensel, 1973).

Although it is now generally agreed that the threshold conditions for temperature sensations can be described by the absolute temperature of the skin at the site of the receptors, the rate of the temperature change, and the area of stimulation, the net sensation can be situation-specific. Learning can alter these values even more.

The Thermal Mesh of the Body Locus

The experimental isolation and stimulation of warm or cold receptors within and beneath the human skin has been infrequent. From the available literature, it appears that the warm and cold receptors are served by unmyelinated nerve fibers. Until recently, unmyelinated fibers were almost impossible to detect physiologically or histologically.

Input of thermal information to the brain follows both discrete and nondiscrete pathways. This constant characteristic of the nervous system has great survival value, although the young neuroscience student belabored with terms and circuitry may question this comment. The most common and well known pathway from the skin to the hypothalamus is through the lateral spinothalamic tract that carries information from the spinal segments (receiving impulses from the skin) to the thalamus. From there, information is sent to the hypothalamus and the cerebral cortex. Thermal sensation for the human face involves another pathway, but still terminates in the thalamus.

Excessive firing of cold receptors and/or the inhibition of warm receptor activity reflexively elicits shivering during acute temperature decreases. The pathways for the shivering thermogenesis are relatively well known. Increased numbers of action potentials leave the posterior hypothalamus and run through the midbrain and the pons. Shivering is controlled via the motor system, involving a tract moving from the cerebrum to the spinal segments and from the reticular nucleus in the midbrain to the spine. Direct cooling of the spinal cord can initiate shivering.

Nonshivering thermogenesis is an effective means of heat production in the neonates of a number of mammalian species, including the human infant. In the adult, this mode is evoked after maintained cold exposure. Nonshivering thermogenesis is mediated by the sympathetic nervous system. This form of thermogenesis is a function of both the skin temperature and the anterior hypothalamic temperature. Shivering from external cooling is depressed when the hypothalamus is maintained at core temperature.

When the warm receptors are stimulated excessively, the primary consequence involves alterations in blood flow and the production of sweat. The consequent evaporative cooling reduces peripheral blood temperature, which ultimately influences the thermosensitive cells in the hypothalamus. Factors that interfere with such cooling, such as copious concentrations of subcutaneous adipose (fat), can drastically alter this mechanism.

The blood flow control within the body boundary varies with the body part. At least three functionally different regions can be distinguished in terms of neurobiochemical mechanisms: the extremities (hands, feet, ears, lips, nose), trunk and proximal limbs, and head and brow. Whereas blood flow through the extremities is controlled exclusively by noradrenergic sympathetic fibers, in the trunk and proximal limbs the mechanism is cholinergic (parasympathetic) via the release of bradykinin from the sweat glands. The head and brow share other mechanisms.

The distinction between the periphery and the trunk has clear advantages for control of energy conservation. This arrangement allows specific alterations of blood flow—and alterations in heat exchange in the limbs—without necessarily involving the central hypothalamic mechanisms. Guyton (1971) points out that very pronounced changes in blood flow can occur locally in the periphery without altering central processes significantly.

When the whole body is exposed to ambient heat, sweating appears on the lower extremities and then proceeds toward the head. Heating of the upper regions decreases the time for whole-body sweating to peak. A particularly high sweating rate is found along the forehead. This quick dissipation option is important since human brain temperature is a function of central blood temperature (unlike many species, such as the rat and goat).

Eccrine thermal sweating in humans is controlled via a relatively odd combination of cholinergic mechanisms within sympathetic pathways. Hensel (1973) concludes that there are only cholinergic fibers innervating the sweat glands, although adrenergic substances can influence sweating. Substances that block acetylcholine transmission, such as atropine, can block neural transmission to the sweat glands.

The General Theory

Most theorists have emphasized the hypothalamus, in particular its preoptic-anterior portion (POAH), as the thermoregulator of the human body. No doubt, there is much evidence to support this proposition. However, to be inclusive, one must realize that the same thermorelated responses evoked by the hypothalamus can be evoked by local stimulation of the spinal cord and thermal responses can be evoked by stimulation of places in the cerebral cortex and the midbrain. There are great interspecies differences.

According to Hensel, electrode (single-cell) measurements of the POAH demonstrated that 60 percent of the tested cells were not temperature-responsive while 10 percent were cold-sensitive and 30 percent were warm-sensitive. The activity of these cells was not homogeneous within temperature ranges. Whereas some cells showed linear increases between temperature change and electrical activity, other cells showed more complex patterns.

Some of the thermosensitive neurons showed a linear or continuous relationship between the induced temperature change and firing rate over a range of from 8°C to 10°C. The most sensitive units often responded with an exponential slope, an interesting observation in light of the hyperbolic relationship between peripheral skin temperature and core temperature. Still other units showed noncontinuous or nonlinear changes whereby thermosensitivity would occur suddenly and then disappear within a part of the temperature range.

Such electrophysiological organization in other parts of the brain has been paired with a type of intensity and intensity-change function. With such constructions, any small change in core temperature cannot stimulate all of the thermosensors. In fact, stimulation of the entire population of thermosensors would induce drastic all-or-none responses.

The existence of thermosensors that respond to the midrange of the temperature alteration allows the existence of specific response profiles. Whereas thermosensors that respond between a 1°C and 2°C change could stimulate one pattern of neuronal activity and behavior, thermosensors that maximally respond to from 3°C to 4°C changes could potentially stimulate another pattern of neuronal activity.

Such neuroelectrical organizations would allow the detection of the total amount of change in core temperature as well as the rate of the temperature change. During extreme temperature changes, statistically more of the thermosensors would be recruited.

The actual trigger of the human hypothalamic response to temperature changes has been assumed to be induced directly. Most theorists assume that caloric alterations of blood temperature from the periphery compose the primary trigger. Interestingly, however, very similar hypothalamic responses would be expected if the weather-related temperature merely desynchronized the normal from 1°C to 2°C change in core temperature that occurs every 24 hours.

This supposition requires much less energetic contribution from the outside environment. The relatively small changes in weather-related temperature would disrupt the normal decreased core temperature during the late night/early morning hours and the increased temperatures of the day.

Only the time, not the magnitude of the POAH activity, would be influenced by warm or cold air masses. Once the air mass stimuli were removed, or following habituation, the normal diurnal variations in core temperature would return. The disruptions would always be temporary and reversible.

Experimental cooling of the POAH in nonhuman primates elicits changes predictable from cold-air exposure. Preoptic cooling by only 2° C, comparable to the daily diurnal range, inhibited the release of vasopressin (ADH) and increased diuresis or urination. Hayward (1972) speculates, with strong data support, that the initial steps of diuretic water elimination occur by shifting the blood to central venous sites. The consequent stimulation of the vagus nerve by swelling volume receptors exhibits a direct inhibition of ADH release. In this general manner, the excess water is shifted out of the blood.

Cooling of the POAH also produces increased EEG activity, especially fast-wave (indicative of arousal); increased body movements; cutaneous vasoconstriction; increased arterial blood pressure; and increased arterial blood temperature. Humoral responses include enhanced secretion of ACTH-cortisone, norepinephrine-epinephrine, and thyroid-stimulating hormone (TSH). Heating of the POAH, in general, produces the opposite effect, particularly the retention of water and decreased diuresis.

The Limbic Connection

One of the most fascinating speculations concerning the development of consciousness in the human being is that our early ancestors first showed transient bouts of consciousness during periods of intense emotional arousal. When these periods of intense alarm attenuated, the fragile condition would slowly dissipate. Presumably, such intervals of consciousness were triggered by the increased number of complex patterns of neuronal firing.

Classic neurophysiologists often parrot the phrase, "The hypothalamus is associated with emotional expression, while emotional experience is associated with the limbic system and its connection to the frontal cortex." Traditionally, expression and experience are coupled together so that the occurrence of one is associated with the occurrence of the other. One experiences emotional responses.

However, there is a fine increment between the occurrence of the hypothalamic response and the occurrence of the "experience." In more pathological cases, such as schizophrenia, the person may show an uncoupling of the two systems. Large portions of his autonomic nervous system may be stimulated—resulting in crying, excessive heart beat, gastric motility, and shivering—yet he will not "experience" these changes or interpret them in an unusual or bizarre manner.

The rich neuroanatomical connections, both known and implied, between the limbic structures and the hypothalamus suggest their interdependence. The hippocampus, cingulate gyrus, septum, amygdala, and some brain-stem correlates (such as the vagus nucleus and solitary tract complex) are known to elicit even the release of ADH and oxytocin from the hypothalamus.

The amygdala and the hippocampus deserve special attention here. The amygdala may not only send but may also receive fibers from a multitude of different hypothalamic nuclei, including the POAH. Experimental electrical stimulation of the amygdala triggers not only ADH release but also apnea, hypoxia, tachycardia, and general autonomic arousal. A few studies indicate a high correlation between amygdaloid function and TSH and ACTH release from the pituitary.

The behavioral correlates of amygdaloid manipulation are extensive and complex. Stimulation or removal of parts of the amygdala in nonhuman primates is associated with fear reactions, deterioration of social interaction, hypersexuality, increased eating or placement of objects in the mouth, and alterations in aggression. Unfortunately, whether the aggression increases or decreases appears to be determined by very subtle specifics.

Alteration in amygdaloid function, according to Hayward (1972), is influenced significantly by blood temperature. In the primate brain, temperature of the blood flowing through the carotid artery is quickly reflected in the temperature of the blood flowing through the Circle of Willis in the ventral part of the brain. Cerebral arterial blood temperature appears to be a major factor in determining the shifts in brain temperature. Very small changes in blood temperature—certainly within the 2°C range of normal variation—can alter amygdaloid function.

According to MacLean (1970), who summarized limbic function in a very general manner, the amygdala is primarily concerned with feelings, emotions, and behaviors that ensure self-preservation. For example, such emotions may be associated with old songs from adolescence or the rich smells of mother's cooking. They recruit intense, sometimes incapacitating, sensations of meaning, vivid images, and a transient dissociation when the person reports, "The present fades into the past."

The most frequent emotional experience reported in surgery during direct stimulation of the amygdala is fear. Epileptic patients, both those with and without motor symptoms, reported feelings of terror, foreboding, sudden strangeness or sudden familiarity, wanting to be alone, and sadness when abnormal electrical activity occurred in the amygdaloid (temporal lobe) region.

Excessive stimulation of the amygdala can precipitate aggression toward the person herself or toward some other object. Since the affective component of meaning is so important to the person's self-concept and interrelation with other people, the fine tuning of this structure can be critical. Delgado (1970) found that assaultive patients could be induced to display intense bouts of intractable aggression following very low-level current stimulation of the amygdala.

In contrast, the septal region appears to be more highly associated with feelings conducive to sociability, procreation, and survival of the species. Stimulation of the septum and other limbic structures can produce immediate ideation of sexual content. The person may suddenly change from a relatively mundane conversation to sentences full of sexual inuendos and seduction. Intense images of sexual behaviors are reported.

The hippocampus is closely associated with amygdaloid functions. Regions of the brain around the human hippocampus have undergone relatively peculiar alterations. For example, the fusiform gyrus, which is folded against the hippocampal gyrus, represents a new convolution in the primate brain. Because of its peculiar blood supply and location in the cranium, the hippocampus has long been recognized as especially vulnerable to mechanical damage, very small vascular insufficiencies, and infection. MacLean (1970) argues that it is the most unstable structure of the brain and has the lowest seizure threshold (without obvious motor movements).

Experimental stimulations of the human hippocampus produce a number of altered sensations, dreamlike states, and feelings of unreality. MacLean argues that enhanced day-to-day bombardment of this region by impulses initiated by autonomic stimuli can enhance emotional startle and alarm from objects entering the peripheral visual field (corner of the eye) and can increase the persistence of paranoid thinking.

Patients who have displayed violent criminal behaviors, such as assault or homicide, often report a sensation of unreality about the situation. They recall a sensation of depersonalization whereby, "Someone else pulled the trigger" or, "Suddenly, without thinking, the person was dead in front of me." Time may seem to slow down or speed up.

Alternatively, they may show perseverant types of private behaviors. During excessive stimulation or chemical alteration, people may report sensations that they cannot move, are fixed in space, or have lost their volition. The sensation is reported as a "dreamlike condition." For example, the person may report that while driving a car, he felt "helpless, detached, and compelled to continually press the gas pedal more and more—as if someone else were in control."

The severity of the difficulties in concentration and problems in retention associated with hippocampal activity would be statistical. Whereas minor changes would produce only occasional distortions, more intense stimulation—primarily from unconscious autonomic stimuli—would enhance the changes to detectable levels. The person may wonder why she spent the entire day in an aggressive, angry, or "just not there" mood. Still other times, the day may be packed with euphoria and glee bordering upon delusion.

The Limbic-Prefrontal Cortex Interface

The conspicuous proliferation of the prefrontal cortex in humans has generated many theories and experiments. This brain structure is associated with the general ability to anticipate and to inhibit responses. Both anticipation and inhibition are tied heavily to the increased memory capacity associated with the large cortical areas.

Anticipation allows people to use subtle cues associated with previous reinforcement and symbols that imply association with reinforcement to modify their behavior before the presentation of rewarding or punishing events. Between the previous reinforcement and the actual delivery of the next, people report experiences of time flow.

The anticipatory capacity is a two-edged sword, however. When the anticipated stimulus is positive, very productive behaviors can be generated. When the expected stimulus is negative, even though the cues for the anticipation may not be conscious, the anticipation generates anxiety. Humans, as a result, compose a species that can be anxious for a lifetime since everyone faces the ultimate aversive stimulus: personal death.

The second basic characteristic of the prefrontal cortexes, inhibition, allows the existence of society. Under normal conditions, various egocentric impulses, deprivation conditions, and limbic behaviors (sexual and aggressive impulses) are inhibited. Complex society operates on the assumption that its members can display a complex series of inhibitory behaviors daily.

Modern Western society is a confusing field of "don'ts." The twentieth century person must learn hundreds of inhibitory responses. He must stop at red lights. She must pass on the left side of the road. He must not be late. She must not tell people what she thinks of them. He must be nice to people higher in the business hierarchy in order to succeed.

When a person is angry, she must learn not to become physically abusive. Sometimes one must learn not to be verbally abusive. When one is tired, one must inhibit the nonsocial behaviors. If one is sexually deprived, one must proceed through expected behavioral patterns in order to obtain reinforcement. Immediate gratification must be inhibited.

When the activity of the prefrontal cortex has been disordered by excessive fatigue or drug intoxication, the general ability to inhibit irrelevant, suppressed, and punished responses is reduced. During marijuana intoxication, for example, the person loses the ability to inhibit irrelevant thoughts and consequently, begins to free associate.

Since the most frequently suppressed (by anxiety or anticipation of punishment) or repressed (by fear of punishment) behaviors are sexual and aggressive responses in Western society, these behaviors become more probable during periods of low-level disruption in the prefrontal cortex. Common drugs, such as alcohol, contribute to this disruption.

During these periods of transient limbic dominance, the person reports being overwhelmed by impulses. Realistic appraisal of the person's situation deteriorates. He may suddenly feel justified in assaulting the person in front of him. She may feel convinced that an entire group of drivers is taking advantage of her and may turn recklessly into the flow of traffic. Guilt and other anticipations of punishment follow limbic activation.

Although the prefrontal cortex usually inhibits these responses, excessive stimulation of limbic pathways from meteorological, social, and disease-related stimuli can overwhelm cortical operation. The first signs are occasional, unwanted impulses or thoughts that "pop" into consciousness.

NONTHERMAL STIMULI

Considering the interdependent nature of the hypothalamic nuclei, and their connections, one might quickly conclude that temperature changes could accommodate all of the changes associated with weather. The data just presented indicate that water balance, vascular changes, hormonal levels, and a variety of moods—from hunger sensations to feelings of self-worth—can be influenced by temperature-triggered changes in the hypothalamus and limbic systems.

There is still reasonable doubt that other components of the weather matrix can contribute separately to the total biobehavioral impact. Teng and Heyer (1955) reported that acute myocardial infarction increased during sudden influxes of cold or warm air masses. More popular allusions refer to the sensations of pain a few hours before the obvious cues associated with a cold front or rain.

Although these reports could be explained by learning factors in which temperature were the initial unconditioned stimulus, each candidate should be systematically evaluated to determine its potential. In science, the acceptance of an effect when it is not real is as undesirable as the rejection of an effect when it is real.

7

MECHANICAL STIMULI OF THE WEATHER MATRIX: BAROMETRIC PRESSURE, WIND, AND INFRASOUND

The weather matrix exerts varied forms of mechanical energy upon the human body. These energies are usually mediated through alterations in the relative pressure of air molecules against body surfaces. Whereas the directions of these forces vary over a three-dimensional hemisphere (with its flat surface against the ground), the amplitude of the greatest pressures can be at least one billion times greater than the smallest effective changes.

The time factor, the duration between successive peaks or lows in pressure fluctuation, is an important differentiating property of these forces. Ultralow frequency ($\sim 10^{-5}$ Hz) alterations in pressure over periods of from hours to days typify barometric changes, while extremely low frequency ($\sim 10^{-2}$ Hz to 10 Hz) air pressure changes over periods of from seconds to about one-tenth of a second typify wind fluctuations and nonaudible sound. Higher and less intense pressure alterations ($\sim 10^2$ Hz to 10^4 Hz) with periods of from 0.01 seconds to less than a millisecond characterize sound. Whereas the mechanisms by which higher frequency pressure alterations influence the human being are relatively well known, the biomechanisms responsive to lower frequency changes are unclear.

BAROMETRIC PRESSURE

The force exerted upon any given point on the earth by the overlying atmosphere has been called barometric pressure (force/area) or atmosphere pressure. A normal atmosphere or simply one atmosphere has been defined as 1,013 millibars at sea level at a latitude of 45°. This value is equal to 29.92 inches or 760 mm of mercury,

406.8 inches (33.9 feet) of water, 1.013×10^6 dynes/cm^2, 1.013×10^5 newtons/m^2, 14.7 pound per square inch (lbs/in^2) or 101.3 kilo-Pascals. A change in pressure of 1 millibar is equal to 0.75 mm = 0.03 inches of mercury = 0.015 lb/in^2.

The multitude of different units reflects both the historical and academic differences in measuring pressure. Length units (inches, millimeters, centimeters) of barometric pressure are direct observations from the length of a column of fluid, such as mercury or water, supported by air. Force units, such as dynes (CGS system), newtons (MKS system), and pounds (English system), applied to area (cm^2, m^2, and in^2, respectively), are convenient measurements for physicists and allow swift incorporation of atmospheric values into routine calculations. The bar, millibar, and kiloPascal are pressure units in which force and area are incorporated.

Atmospheric pressure is influenced by temperature, altitude, latitude, and water vapor. Variations in temperature alter the average distance between air molecules. Air molecules are closer together and exert more force over an area in cold air than in warm air. Variations in altitude determine the absolute number of air molecules present. As the altitude increases, the density and, hence, the force exerted by air decreases. Pressure decreases by approximately one-thirtieth of its value with an increase of 275 m (900 feet). Whereas normal sea level pressure is 1,013 millibars atmospheric pressure, atmospheric pressure 275 meters higher would be approximately 979 millibars.

Latitude variations in atmospheric pressure are primarily a consequence of gravity. The effects of gravity are greater upon the air overlying the poles (high latitudes) than upon the air over the equator (low latitudes) where the centrifugal force of rotation compensates somewhat for gravity. The relative concentration of water vapor, which can vary from almost 0 percent to about 4 percent of the air's composition, becomes a significant factor when precise calculations are required.

In order to compare barometric readings across different points in terrestrial space, adjustments are routinely made for temperature, latitude, altitude, and instrumental factors. Although a necessary procedure for demonstration of meteorological phenomena, it should be remembered that adjusted values may mask the absolute values to which people are exposed.

The extremes of atmospheric pressure adjusted to normal atmosphere have been recorded in Weatherwise (June 1971, p. 130-31). According to this record, the highest barometric pressure readings in the world, Canada, and the United States were 1,084 mb (Siberia, December 1968), 1,067 mb (Medicine Hat, Alberta, January 1897), and 1,063 mb (Helena, Montana, January 1962), respectively. The lowest barometric pressure measurements in the world, Canada, and the United States have been 877 mb (19°N 135°E, September 1958), 946 mb (Gander, Newfoundland, February 1964), and 892 mb (Matecumbe Key, Florida, September 1935), respectively. Such extremes are associated with severe weather conditions.

Temporal Variations

The absolute change in atmospheric pressure associated with the weather matrix at any given altitude and latitude is only from about 1 percent to 4 percent of the total value. Typical (that is, once every two weeks) net variations in barometric pressure associated with low pressure (cyclonic) air masses range from 40 mb to 50 mb in Europe. Higher latitudes are prone to greater extremes of up to 120 mb, while equatorial regions display relatively small variations of 20 mb or less. The majority of weather changes in North America and central Europe are associated with atmospheric pressure changes less than 25 mb.

Atmospheric pressure changes associated with the weather matrix occur within time intervals of from 10 hours to 100 hours. As can be seen in Figure 14, the number of comparable changes within, for example, a seven-day period are quite variable. Some weeks are associated with very little alteration in atmospheric pressure (for example, stationary high pressure masses), while other weeks are associated with two or three distinct variations. To some extent, the numbers of air masses and pressure variations per week period are influenced by season.

The temporal range (from 0.5 days to 4.0 days) in comparable pressure changes indicates the clear existence of a derivative function in this phenomenon: rate of change. A 30 mb change within a 12-hour period would occur at least 8 times more quickly than a 30 mb change with a 4-day (96-hour) period. In context of the homeostatic model of weather response, different rates of change would elicit very different patterns in human response. Such temporal distinctions could be the difference between a steplike and ramplike signal.

Assuming a linear (straightline) decrease or increase in atmospheric pressure associated with an approaching air mass, a 30 mb change in 12 hours would show a rate of change equal to 2.5 mb/hour, while a comparable change in 72 hours would be 0.4 mb/hour. However, the rate of change (see Figure 14) is not always linear or equal. Some low pressure areas can be associated with accelerating decreases in barometric pressure during the latter stages of approach. At the point of inflection on the curve, the rate of change could be

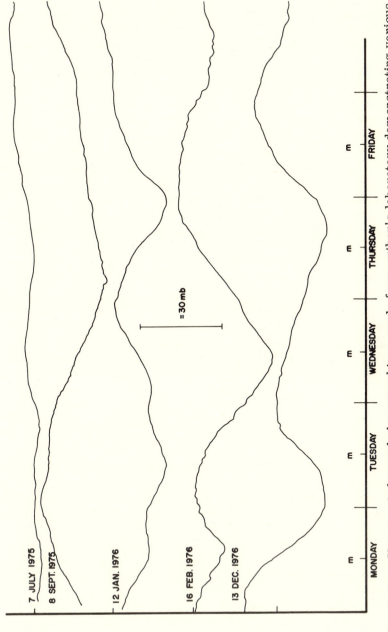

Figure 14   Uncorrected sample barographic records from author's laboratory demonstrating various profiles of change.   Amplitude is indicated by vertical line (30 mb).   Time increments are 24 hours (m = midnight).

quite high. Such values would still be quite low compared with the minimum of from 2 mb to 3 mb/second change (regardless of direction) required for reliable detection by human subjects (Williams & Cohen, 1972).

Thunderstorms and related local disturbances can generate variations for brief periods. On an absolute level, these changes are very small, existing as "little squiggles" superimposed upon the much higher and longer daily or weekly pressure waves. Net changes in one direction usually range from 1 mb to 10 mb within 20 minutes to 60 minutes. The rate of change, however, is quite great. When adjusted to hourly intervals, rates of pressure change can approach 20 mb/hour. Although almost an order of magnitude ($\times$ 10) larger than change rates associated with larger air masses, these values seldom exist for more than one hour.

Still smaller variations in atmospheric pressure are associated with the earth's rotation and with lunar tides. Daily variations with two maximums (10 a.m. and 10 p.m. local time) and two minimums (about 4 a.m. and 4 p.m. local time) are related in a complex manner to the gravitational attraction of the sun and the daily changes in temperature. The amplitude of this variation is about 3 mb at the equator and decreases toward the poles. The amplitude of the lunar tides is about 0.1 mb per 24 hours. A comparable change in pressure can be achieved by standing on a stepladder one meter high.

The consequences of barometric pressure change have been studied through both experimental and correlational techniques. Although experimental studies indicate changes in the order of two to three millibars within about a second for clear detection, the distinction between detection and effect must be emphasized. Many different environmental effects can exert changes that are displayed at nonaware levels of human behavior. Psychological detection is not necessarily a requirement for biobehavioral effect.

Sample Correlational Studies

Decreases in atmospheric pressure of approximately 40 mb/24 hours have been associated with increased restlessness in populations of schizophrenic patients, especially females, even though housing temperatures were presumably controlled (Tromp, 1963). Comparable observations of the relatively vague measure "restlessness" and of increased ambulatory behavior have been reported for a variety of nonhuman animals during similar rates of change.

More discrete measures, such as numbers of some specific ailments, allow clearer correlations. The passage of cold fronts, provided these stimuli were separated by several days, have been reported to be associated with decreased fibrin formation in the blood (increased fibrinolysis) and slower clotting time. These correlations suggested that diseases associated with this activity, peripheral arterial embolism and peptic ulceration, should be correlated with changes in barometric pressure.

J. B. Hansen (1970) correlated 140 cases of peripheral arterial embolism (PAE) with various barometric pressure measures on the day of the attack and from the previous two days. This disorder, characterized by either a sudden onset of pain or a slow increase in pain over a few hours in the affected limb, is primarily due to clot formation in an artery of the leg or arm. Hansen compared the frequency of onset according to mean barometric pressure, mean rate of change over the three-day period, the range (difference between highest and lowest value), and the maximum rate of change per hour.

The percentages of patients displaying the PAE disorder as a function of the six intervals of maximum rate of change are shown in Figure 15. Hansen's results clearly indicate a linear relationship between the risk of PAE and the maximum rate of barometric pressure change during the previous two to three days. Such significant and clear relationships were not found for any of the other barometric pressure measurements. Close inspection of the data demonstrate that a major component of the significant effect was contributed by marked increased in PAE responses for the from 0.4 mb/hr to 0.6 mb/hr changes.

A second pressure-human ailment correlation was reported by Hansen and Pedersen (1972). In this case, 190 male and female patients treated for perforated duodenal ulceration were selected as subjects. Using the same four measures of barometric pressure, numbers of ulceration onsets were investigated. They found more cases than expected by chance during total changes of more than 8 mb than with lower changes. Again, the strongest relationship occurred between perforation frequency and maximum rate of change. In this study, a from 0.5 mb/hr to 0.6 mb/hr change was associated with the greatest incidence above expectancy values.

Critique of Correlational Studies

In addition to the general problems of correlational analyses (Chapter 5), barometric pressure changes in these disorders are confounded by other changes in the triad. Development of cardio-vascular disorders is correlated not only with atmospheric pressure alterations but also with changes in air temperature and humidity. Usually medical records do not specify how frequently or how long

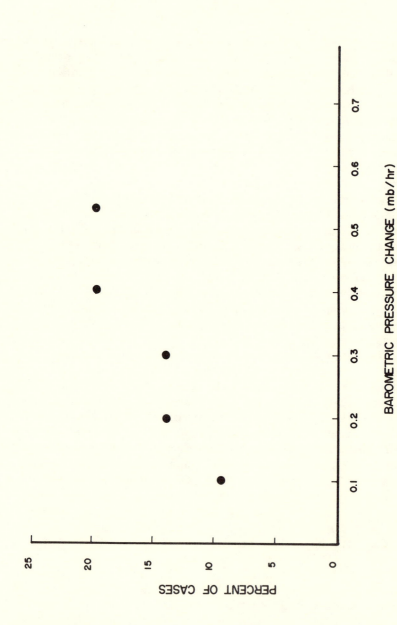

Figure 15   Percent of peripheral arterial embolism cases reported as a function of change in barometric pressure (mb/hr). Data modified from Hansen, 1970.

patients prone to weather disorders were exposed outside their dwellings to these factors before the disorder became apparent.

The failure to differentiate the direction of change is a second
limitation of the Hansen correlations. No doubt, Hansen and his
colleagues have clearly demonstrated the importance of rate of change
in the development of pathology. Whereas increasing rates of change
might be associated with cold air masses, decreasing rates of change
might be associated with warmer air masses. Although low pressures
appear to be associated with more water retention than high pressures,
air temperature can modify this response.

Since Hansen and others rarely report rates of change in other
parameters, the possibility of a confounding factor or interaction
from other parts of the weather matrix cannot be eliminated, even
partially. The reader does not know how frequently air masses
moved over the population studied. Sensitive populations who have
not been exposed to a major change for two or more weeks would be
expected to display more vigorous symptoms than a population that
has been inundated with different pressure changes within a single
week.

The Hollander Experiments

Several experimenters have noted increased water retention
in nonhuman animals associated with drops in atmospheric pressures
of from 35 mb to 130 mb per from 12 to 48 hours. Usually water
retention correlated with this stimulus is also associated with increased "restlessness" in the experimental animals. Such measures
are crude and often laboratory-specific. Controlled experiments
with human subjects have rarely been completed, primarily due to
the monetary and ethical restraints of human experimentation.

One of the most well known series of controlled experiments
with human subjects was reported by J. L. Hollander (1963), who
tested the frequent complaint that weather changes precede joint
discomfort. Hollander and his colleagues designed and constructed
a special "climatron" or controlled climate chamber within which
temperature, humidity, atmospheric pressure, and ionization levels
could be controlled. Rates of change could be controlled as well.

Weather-sensitive patients diagnosed with rheumatoid arthritis
or osteoarthritis were selected to live inside the chamber for from
two to three weeks. The chamber was constructed so that a variety
of intervening psychological variables were attenuated. Patients
were visited by friends and relatives and had direct access to television and radio. A window in the chamber provided a very important
option for patients involved with isolation or compartment experiments:

a contact with the outside world. Many of the patients reported that they had never "had it so good."

During the first five to seven days of each experiment, the patients were habituated to a constant environment 76° F (24° C), 40 percent humidity, 1,016 mb pressure, 400-600 negative ions/ml^3 of air, and 200-300 positive ions/ml^3 of air. The rate of air flow was maintained at 305 m^3 (1,000 ft^3)/minute. Subjects were instructed to record their experiences in diaries throughout the experimental period. Other measurements of arthritic symptoms were made by attending staff.

The subjects appeared to have adapted well to the climatron. Except for the first two days of exposure, when many subjects reported temperature changes even though they did not occur, subjective reports remained relatively constant. Even though patients could see the outside world, observation of rain or cloudy skies did not appear to influence the subjective experiences. Hollander reported that many of his patients expressed "wonder" at the lack of such effects, such as, "It's raining today and I didn't feel it coming."

Without the patient's knowledge, the four different weather factors were manipulated singly and in combination. Barometric pressure was raised steadily for periods of from 4 hours to 24 hours up to 1,084 mb, then returned to 1,016 mb. The range in temperature was from 60°F to 90°F, while the range in humidity was from 30 percent to 80 percent over similar periods. Positive ions were increased to 60,000/ml^3, and negative ions were increased to 50,000/ml^3. During the 60 different trials involving 24 different subjects, these variables were randomly presented in various combinations.

Hollander reported that merely dropping the temperature was associated with increased stiffness in five cases. The evoked sensation of chilliness was rectified quickly when the subjects put on sweaters or coats. At constant temperature (76°F), markedly increased humidity was associated with reported sensations of coldness. There were only three instances of subjective worsening with humidity changes.

The most significant and reliable worsening of both subjective reports and clinical evaluations occurred with a simultaneous drop in barometric pressure and a rise in relative humidity. The subjective symptoms increased within four hours of the simultaneous change and were very obvious within from six to eight hours. According to Hollander, these effects were evident in 29 of the 46 trials in which this combination of variables was used.

If the humidity were kept constant for a day at the peak of 80 percent and the barometric pressure were maintained at the low of 965 mb, complaints decreased and symptoms improved. On the other hand, if the variation cycle of these two factors were repeated daily, the conditions and symptoms of the patients became progressively worse. Collectively, these data demonstrated that the change in the two factors, not their absolute or steady state values per se, were the critical contributors to the symptoms measured.

Assuming that the changes in barometric pressure and humidity were constant (sine-wavelike), the drop in pressure ranged from 2 mb/hour for 24 hours to 17 mb/hr for 4 hours, while the increase in humidity ranged from about 2 percent/hour (over 24 hours) to 10 percent/hour (over four hours). These values are quite comparable with the short-term changes associated with thunderstorms and with the slower rates associated with major air masses.

Certainly, Hollander's experiments were impressive, considering the decade in which they were completed and the nature of the animal studied. However, they are not conclusive. For example, not all subjects responded to the increasing humidity/decreasing barometric pressure change. Although the average increased severity of arthritic complaints was 63 percent, the range was from 0 percent to 100 percent. One of the eight subjects in one study reported no effect, while two of the subjects always showed an effect. The others were inconsistent.

There are several criticisms of the Hollander data and methodology. His procedures may have masked arthritic effects since he frequently instituted the barometric pressure-humidity changes everyday for up to five days in one week. Secondly, he frequently presented the changes while the subject was still reporting complaints presumably associated with a previous change.

The most serious design problem in Hollander's experiment, indeed within any experiment requiring human subjects, is the lack of control against experimenter bias. He does not report controls against subtle communication of the environmental condition by himself or his support staff during the varied and unspecified forms of communication with the patients. Facial cues and voice intonations, especially from individuals who control a person's environment (from boss to parent), can be potent sources of information. Comments such as "The patient's responses were not due to psychological effects since they did not respond to cloudy skies," is an inference, not a data statement. They may have been responding to the complex social demand characteristics of the situation. The persistent and potent contribution of psychological factors to behaviors highly correlated with weather changes must never be forgotten.

A General Mechanism

Why should the simultaneous change in barometric pressure and humidity be a critical factor in arthritic complaints? What is the importance of the rate change? One general mechanism that is a summary of several specific hypotheses involves a barometric pressure change-related alteration in body fluids. It depends heavily upon the two modes of removing excess water: passively, by evaporation through the lungs and skin, and actively, through the kidneys. The general problem would involve differential lag latencies.

How would this work? First, assume there are two functional systems within the human body that respond to changes, or to their consequences, in barometric pressure. Anatomically, these systems may not be distinct, but may be merely two hierarchical aspects of the same system. One system responds to rates of change by maintaining water retention. The second system responds to absolute change only after a sufficient number of intervals have occurred to comprise a step function. Whereas the first responds in latencies of from minutes to hours, the second requires from hours to a day or so.

As the barometric pressure falls, body space expands very, very slightly, stimulating receptors responsive to water volume and water pressure (osmolarity). More water is retained. As the water accumulation slowly increases (ramp function), the relative accumulation increases within the extracellular space. Areas that preferentially maintain water would be statistically more affected by this slow accumulation.

Parts of the body containing large amounts of proteinpolysaccharides would be affected significantly by this slight disparity in water balance. One would predict that the synovial volume of the joints would increase, causing swelling and stimulation of pain receptors in the vicinity. Any area with a disproportionate increase in these proteinpolysaccharides, such as healed bone fractures and surgical scars, would gain relatively more water, swell, and stimulate local pain receptors.

Patients with eye diseases, such as glaucoma, would report enhanced pain and swelling due to the high concentration of hyaluronate within the eye humors. Depending upon heart history, similar increased water concentration in the heart regions would influence blood pressure. The effects of the transient storage of water on blood clotting, via heparin, would be expected. Since the half-time for clotting is so quick relative to the duration of the barometric stimulus, overshoots and undershoots (excessive clot formation and bleeding) would be expected.

The brain, which works on electrical digital codes, would be influenced as well. Increased extracellular fluid would allow a few potassium ions to flow out from the neuron and a few sodium ions to flow into the neuron. The net effect would be to place the neuron closer to firing (more irritable). When it did fire, the height of the action potential (its clarity) would be reduced. Although the neurons would be stimulated more easily by sensory input, the number of inputs must be repeated in order to obtain the same resolution. Grossly, the person might report more free-associative thinking and greater difficulty in concentrating.

The humidity factor would be an expected and critical aspect of the barometric change phenomenon. The primary nonkidney modes of eliminating water from the body are through the skin and lungs. If the outside air were dry, any slowly accumulated water would be expelled easily by evaporation. However, if the outside air were becoming more humid, the water accumulating in response to the barometric pressure change could not be eliminated easily. Consequently, the water would be maintained within the body and allocated to the storage areas previously mentioned.

Once the accumulation of water achieved a critical level, probably mediated through an electrolyte concentration sensor, the second control mechanism would be stimulated. This mechanism would be more powerful and higher on the organismic hierarchy of homeostats than the initial system. Once the more powerful system is stimulated, the pressure-rate-sensitive system of water accumulation would be terminated.

The consequent overriding responses would involve removal of excess fluids through the kidneys, stimulation of the hypothalamic control centers, and general activity of the adrenal gland. Adrenal activity would release electrolyte-controlling chemicals and anti-swelling (anti-inflammatory) chemicals into the blood. Relative electrolyte balance would return to normal, and the symptoms would habituate slowly to a baseline level.

The general concept of weather-related complaints of this type would be a very oversensitive low-level control system that is capable of slowly initiating disequilibrium before the severity of the change is detected by the more powerful homeostatic control. The latter system must be relatively insensitive, however, otherwise it would be stimulated by every trivial change in the environment.

The reader is reminded that the previous discussion is a story dressed in scientific terminology. In principle, even the most complex biobehavioral changes associated with the weather matrix can be described or explained. There is a very large gap, however, between the false sense of prediction afforded by semantics and the specific, quantitative, testable requirements of a theory.

WIND

Wind is air moving in the horizontal plane. In many respects, wind contributes most to weather since it is air in motion. Masses

of warm air move quickly over large areas toward regions of colder
air and masses of cold air move to warmer regions; the relatively
motionless human being is exposed to this movement. If air has
moved over cold and snow-covered regions, it will be cold. If air
has moved over oceans and bodies of water, it will be damp.

Characteristics

Wind is a vector quantity; it has both direction of movement
and magnitude. Rarely are these two characteristics of wind consistent over long periods of time. Usually, they fluctuate from slow
to fast or move from one deviation to another. The absence of stream-
line flow is called turbulence. Turbulence is typified by successive
gusts and lulls lasting a few seconds.

Turbulence increases as a function of wind speed. Surface
characteristics of the medium over which air is moving also determine the severity of turbulence. It is greater over land than over
ocean surfaces and greater over forests and cities than over bare
plains. Turbulence is analogous to currents induced along river
banks where the quick flow of the water is inhibited by frictional drag.
Air turbulence extends five to six times higher than the obstacles
that produce it.

Winds vary as a function of height, season, and time of day.
In general, the average wind velocity increases with height above
the ground, especially in the first 30 meters. During some weather
conditions, the wind near the ground may be in the opposite direction
of the wind less than 100 meters above it. Such conditions generate,
not surprisingly, a great deal of turbulence: rapid fluctuations in
air movement.

The velocity of the wind is generally greater during daylight
than at night over land areas. The highest average occurs in the
afternoon, while the lowest occurs around sunrise. This diurnal
variation is related to the heating and rising of surface air by day
and its descent by night. Over water areas, where diurnal temperature variation of air is minimal, there is little difference between
nocturnal and daytime wind velocities.

Differential heating, either between land and water masses or
between upper and lower land masses, can generate winds or breezes.
Along seacoasts, the land warms more during the day than does the
adjacent water. When the warm air expands and rises, the cooler
air from the sea moves inland, creating the sea breeze. The sea
breeze is a shallow air layer (from 200 m to 400 m thick) and extends
to distances of from only 20 km to 50 km inland. Around smaller
bodies of water, such as the Great Lakes, the lake breeze extends
from only 1 km to 5 km inland.

At night, the land cools more than the water, resulting i
density air relative to the air over water. As the air moves ou
sea, the land breeze is created due to the local cooling. The la
breeze is less developed than the sea breeze; it is shallower, ha
less extension seaward, and displays less velocity.

A vertical example of breezes or winds created by differential
heating is valley and mountain breezes. When the floor of a valley
is heated during the day, the hot air rises and results in the movement
of warmed air up the valley or mountainsides. During the evening,
the air cools more quickly on the mountainsides than in the valley.
As the cool air moves under the action of gravity into the valley, the
mountain breeze is created.

Both the direction and speed of wind change annually, depending
upon a number of general circulation and local factors. In general,
the velocity of wind is greater in the winter and spring than in the
summer and autumn. The predominance of the former two seasons
is related, not surprisingly, to the greater differential temperature
contrast between high and low latitudes. Local temperature contrasts
from rivers, mountains, or general topography can alter annual peak
wind times.

Organismic Relevance

Wind is a vector quantity from both physical and biological
perspectives. Movements of air determine the relative concentrations and availability of microorganisms and allergens to people.
Wind carries ions of different size and charge and distorts the local
electric field. Wind pressure exerts its effects upon the whole body,
the eardrum, and the ear canal. By enhancing the effects of humidity
and temperature, wind contributes to the chill factor.

Wind increases the rate at which particular matter, such as
allergens and aerosols, in the air is exposed to the lung linings and
nasal passages of sensitive people. The net effect of increasing the
rate of exposure to atmospheric particles is to increase the total
amount or dosage of the material. Whereas the concentration during
no wind would be subthreshold for allergic response displays, the
wind factor can push the net concentration of breathed allergens to
reactive levels. Such wind factors are more obvious to allergic
people while driving an automobile with the windows open during the
allergy season.

Moving air exerts pressure against objects in its path. Within
normal ranges, wind pressure is equal to the square of the velocity
times a constant. If the speed were expressed in miles per hour
(mph), the constant is 0.004 and the pressure is expressed in pounds

per square foot (lb/ft^2). Wind blowing at 10 mph would generate a
pressure of 0.4 lb/ft^2, while wind blowing at 50 mph would generate
a pressure of 10 lb/ft^2. Expressed in another system, the pressure
generated by 10 mph and 50 mph winds would be about 190 dynes/cm^2
and 4,800 dynes/cm^2, respectively.

Compared with vertical pressure at sea level, which is 2,117
lb/ft^2, wind pressure would be trivial. However, there are significant differences in the application of the two: direction and time.
The human being is constructed mechanically to sustain vertical
pressure more effectively than horizontal pressure. Atmospheric
pressure is applied in a continuous and more or less constant manner
(with 1 percent variations over hours), while wind pressure varies
markedly within very short periods (over seconds). In terms of
sheer energy and power, wind pressure would not be expected to
influence the human system through the same mechanisms that mediate atmospheric pressure changes.

Direct effect of wind upon the eardrum may appear likely at
initial glance. After all, the pressure exerted by wind speeds of
about 20 mph is about 500 dynes/cm^2, and the ear is responsive to
pressures between 10^{-4} dynes/cm^2 and 1,000 dynes/cm^2. However,
again the critical feature is a time factor: the number of oscillations
per unit time. The ear is maximally responsive to air pressure
waves that strike the ear drums at about 1,000 times a second (1 kHz).
At this frequency, the ear is so sensitive that only 0.0002 dynes/cm^2
is required for hearing (Woodworth & Schlosberg, 1962, pp. 323-61).

Since wind typically does not oscillate against the eardrum at
frequencies within hearing range (between 20 Hz and 20,000 Hz),
sound experience should not occur. Sensations of coolness or tactile
experiences from pressure against small pain/touch receptors could
result. Sound experiences may be generated harmonics from air
rushing past protruding portions of the ear canal.

Higher velocity winds add a transient but important factor to
human behavior: noise. Depending upon local shapes of buildings
and local geographical forms, wind or gusts in excess of 20 km/hour
can produce a large range of mechanical distortions in the air that
are perceived as sound. Psychologically, these mechanical waves
within the air medium are experienced as shrills or groans. Prolonged exposure to the changing frequency and intensity of wind noise
has been demonstrated to increase irritability in some people.

Wind chill is another consequence of air speed. In the most
general form, wind chill is a measure of the speed and the amount
by which heat is removed from the body through the skin. The relative effect of wind chill is a function of the temperature, humidity,
and wind speed. Heat loss becomes greater as wind speed increases
and ambient temperature decreases. High humidity contributes to

this loss by reinforcing the effects of heat loss from the skin covering and the lung lining. For example, if the local temperature were -5°C and the wind speed were 40 km/hour, the person would display heat loss as if the temperature were about -25°C.

Even though a person may be inside, moderate wind speeds produce a "home breeze" by facilitating the exchange of air across the walls. Wind pressure applied against the walls of the house increases diffusion of outside gases into the living area. Increases in drafts around windows and along draft pathways contribute to a low-level but behaviorally relevant chill factor within the home environment.

Correlational and Experimental Studies

Despite the early observations by Greek physicians that inhabitants of towns exposed to wind frequently develop more diseases, few studies of wind velocity (as an event) and human populations have been reported. Presumably, according to early observations (many of which were replete with confounding variables), people living in windy places were prone to ulcerations, difficult healing, general lethargy, convulsions, and loss in fertility. More recent experiments with wind speed or air flow have been oriented toward the effects of wind on heat loss.

SPECIAL WINDS

Whereas the "everyday winds" associated with heat exchange between large air masses or cyclonic/anticyclonic conditions have been considered mundane, frequent attention has been oriented toward the so-called special winds. Special winds usually occur suddenly, are seasonal, and evoke marked alterations in local weather. These steplike inputs to human populations have long histories of bad effects.

The most well known special wind is the Foehn, which blows from the south down the northern valleys of the Alps. Some meteorologists suggest that the Foehn starts with the hot winds in the Sahara Desert that move across Italy and then to the Alps. The Foehn usually coincides with low pressure areas over northwestern Switzerland and a storm over the Atlantic. However, Foehn-like winds are likely along the far side of any mountain range where air descends after a forced rise, loss of moisture, and heating.

The Foehn is a hot, dry wind that can occur anytime during the day or night. It is associated with a sudden rise in temperature, in the order of 10°C, and exceptional visibility. As the hot wind continues, fire risks increase to critical levels. People report general discomfort, shorter tempers, and a variety of well known weather-related complaints, from embolisms to traffic accidents.

Other special winds include the Lombarde in Italy and France, the autan that blows over Haut-Languedoc, and the midi that is evident around Lyons, France. The latter is well known to be associated with a sudden rise in temperature and reduction in humidity. Discomfort, headaches, rheumatic pains, stifling sensations, irritability, asthma attacks, suicides, and "electrical feelings" are common correlates.

The best known North American special wind is the Chinook or snow eater. This term has been used to designate the wind blowing down the valleys of the Missouri River tributaries as well as to designate the hot and dry wind descending from the Rocky Mountains. The Chinook has striking effects around Calgary, Alberta, Canada, where the winters can be quite variable. When it blows in the winter, temperatures may rise more than 30°C within a few hours, for instance, from -20°C to +10°C. Compared with the Foehn in Bavaria, the Chinook, although paired with similar complaints, appears less effective. However, no direct study has correlated the length of time these special winds blow with the severity of symptoms. A stimulus that occurs for a long period of time, even with adverse effects, becomes old news quickly; a stimulus that occurs for a short time and then disappears is a discrete clear event that maintains its novelty and its news coverage.

The major components of special winds that lead to their behavioral-biological complaints, assuming they are not determined by tradition and expectancy, have usually been thought to be high temperature and low humidity. Certainly, increased temperature and low humidity would produce undesirable elevations in house temperatures, even when controlled by homeostats adjusted to winter temperatures. Water evaporation from the lung lining and skin would be significant. The enhancement of electric charge effects, from irritating discomfort of touching the ignition key after sliding across the car seat to touching the doorknob after walking across the carpet, are little factors that contribute to the irritability of everyday life.

However, not all special winds share the high temperature/humidity characteristics presumed to be primary correlates of human complaints. The Levanter of Tangiers, an eastern wind that blows through the Straits of Gibraltar, is associated with cooler air and little change in barometric pressure or humidity. However, similar types of complaints have been reported. Using what appears to be "it's the only thing we know about that's left" logic, some theorists have considered electric field variations and ionization as culprits of special wind effects.

Despite the observations of enhanced ulceration, poorer healing, general malaise, and irritability recorded by physicians almost 2,000 years ago, few quantitative correlative studies have been conducted with these winds. Those that have been conducted, especially related to the Foehn, have not differentiated its effects from psychological and social factors. The control factors, no doubt, are many.

INFRASOUND

The term <u>wave</u> in physics refers to a propagational disturbance that occurs when a field fluctuates with time. Waves that involve the displacement of some mass of material from its equilibrium position include sound waves, water waves, and oscillations through solids. The most common atmospheric wave is associated with gas molecules moving back and forth due to some vibrating system, such as a loudspeaker or human voice. When the number of disturbances per unit time (the frequency) and the total displacement of the molecules bouncing back and forth (the intensity) are within a certain range, the experience of sound occurs.

Characteristics of Normal Sound

The experience of sound occurs when pressure oscillations in the air range between 20 Hz and 20,000 Hz. Different oscillations within that range are experienced as changes in pitch or tone. The experience of loudness is a function of how vigorously the molecules move back and forth from equilibrium. At excessive levels of perturbation, the experience of sound changes to pain. A frequent anatomical correlation is destruction of the ear's small hair receptors that change mechanical stimuli to electrical-chemical stimuli: the brain's information mode.

Although the human ear can respond to pressure changes in air between 10^{-4} dynes/cm^2 and 1,000 dynes/cm^2, the minimum intensity for detection of sound changes with frequency. For example, only 10^{-4} dynes/cm^2 is required for sound experience if the air molecules oscillate 1,000 times per second (1,000 Hz). As the frequency increases or decreases, more and more air molecule displacement is required to experience sound. At about 10,000 Hz, pressures of 0.01 dynes/cm^2 are required, while at 100 Hz, about 0.1 dyne/cm^2 is required. For frequencies between 500 Hz and 10,000 Hz, pressures against the eardrum of about 1,000 dynes/cm^2 begin to elicit sensations of pain or unpleasant pressure. Below 500 Hz, the thresholds for pain are slightly higher, leveling off at about

5,000 dynes/cm^2 at the lower end of the hearing range (Woodworth & Schlosberg, 1962).

Sound intensity is reported frequently in <u>decibels</u> (db) rather than in dynes/cm^2. The decibel is a ratio of energy levels. An energy ratio of 10:1 is 10 db, while a ratio of 100:1 is 20 db, while a ratio of 1,000:1 is 30 db, and so forth. Since the log of 10 is 1, the log of 100 is 2, and the log of any number starting with 1 and followed by some zeros is equal to the number of zeros, the db can be easily calculated by multiplying the number of zeros by 10.

The reference level (the "1" in the previous ratios) used by most psychophysicists who study hearing phenomena is 0.0002 dynes/cm^2 (the average human stimulus threshold for a tone of 1,000 Hz). Assuming db = 0 for that value, increases of increments of 20 db are associated with an order of magnitude increase in dynes/cm^2. Consequently, 0.001 dynes/cm^2 = 20 db, while 0.01 dynes/cm^2 = 40 db; 0.1 dynes/cm^2 = 60 db, 1 dyne/cm^2 = 80 db; 10 dynes/cm^2 = 100 db, 100 dynes/cm^2 = 120 db; and so forth. To obtain the energy in watts/cm^2, simply square the pressure measurement; that is, 10^{-2} dyne/cm^2 = 10^{-4} watts/cm^2, 10 dynes/cm^2 = 100 watts/cm^2. Values for sound stimuli in typical audible ranges (100 Hz to 10,000 Hz) would be 40 db for an average office, from 50 db to 60 db (about a milliwatt) for a conversation, 80 db for a pneumatic drill, 110 db for an airplane engine, and more than 120 db for close thunder.

Characteristics of Infrasound

Whereas the term <u>ultrasound</u> refers to mechanical vibrations in matter (especially air) above the normal human range (greater than 20,000 Hz), <u>infrasound</u> refers to air vibrations below the normal hearing range. As the frequency decreases, not surprisingly, the air pressure required to obtain a hearing sensation increases markedly. At 100 Hz, the value is 0.1 dyne/cm^2 (60 db); at 10 Hz, the value is about 100 dynes/cm^2 (120 db), while at 1 Hz, the value appears to be about 1,000 dynes/cm^2 (140 db).

Apparently, the pain thresholds for infrasound levels between 1 Hz and 20 Hz are higher than for more audible frequencies. According to early measurements, "zero" frequency or static (constant) pressures of 179 db (about 100,000 dynes/cm^2) are required before pain is reported by presumably normal people. When the pressure is altered at 3 Hz, the pain threshold drops to 165 db (25,000 dynes/cm^2). Between 15 Hz and 100 Hz, as mentioned, the pain threshold levels off around 140 db (1,000 dynes/cm^2). Some theorists have concluded that infrasound at 200 db (1 million dynes/cm^2) would be deadly, due to excessive vibration and displacement of the internal organs.

As does normal sound, infrasonic waves travel at about 344 m/sec at 20°C. The length of the wave is an inverse function of its frequency. Consequently, a 1 Hz wave is 344 m long (0.344 km), a 0.1 Hz wave is 3.44 km long, and a 10 Hz wave is 34.4 m long. The frequency of a wavelength equal to the height of a 1.86 m (6 ft) man would be about 540 Hz. An infrasonic wave with a length equal to a 100 meter tall building would have a frequency of 3.4 Hz.

Unlike normal sound, these waves can travel very long distances over several hours without appreciable energy loss. These waves have two options of propagation: along a straight eyeline path or within an earth surface/upper atmosphere wave guide. A 1 Hz infrasonic wave, within the optimal propagation mode, can travel 10,000 km and lose only 1 percent of its initial power. A 0.1 Hz wave traveling around the entire earth loses about 5 percent of its initial power. Losses due to scattering by hills, buildings, and local topography would be quite small for wavelengths greater than 1 km.

Sources of Infrasound

Infrasound waves have been correlated with the occurrence of meteors entering the atmosphere, presumed tectonic (structural) strain within mountain regions, volcanic eruptions, geomagnetic storms, tornados, severe weather or storm conditions, and special winds. Amplitudes of infrasound from these sources range from 0.1 dynes/cm^2 to about 50 dynes/cm^2. Different environmental phenomena appear to generate different power peaks in the infrasonic spectrum, with some frequencies less than 0.01 Hz.

Because of their capacity for long distance propagation with minimum energy loss, sources of infrasonic waves are often difficult to determine. Infrasound detection has been associated with low-level earthquake activity (microseisms) more than 200 km from the measurement point and with heights of water waves in the North Atlantic several hundreds of kilometers away. Infrasonic waves between 0.1 Hz and 0.01 Hz with amplitudes between 1 dyne/cm^2 and 10 dynes/cm^2 have been detected thousands of kilometers away from sources within mountain ranges in British Columbia, Canada. These waves display both diurnal and seasonal variations. Even the entrance of intense fireballs (meteors) into the atmosphere can generate comparable infrasonic frequencies and magnitudes.

Singular energetic events from both meteorological and geophysical sources can generate infrasound over tens of thousands of kilometers. Infrasound of less than 1 dyne/cm^2 was found to be correlated with tornado swarms more than 2,000 km away from the detection sight in the United States. According to Cook (1962), infrasonic waves traveled around the earth several times following the massive explosion of Krakatoa in the East Indies in 1883. Geomagnetic storms, especially sudden, intense varieties, are also correlated with increased global infrasonic output for between 1 hour and 24 hours. The source has been estimated to be fixed within geomagnetic coordinates on the side of the earth opposite the sun. Vigorous earthquakes, such as the Chilean quake of 1960, can generate infrasound around the earth.

Weather-related sources of infrasound have received significant attention since the initial works of Gavreau (1968). During severe storms, infrasound between 1 Hz and 4 Hz may reach 110 db (Brown, 1973). A full-force gale can generate infrasound around 1 Hz at 130 db. Special winds, such as the Ministrel in Southern France, are associated with comparable amplitudes of infrasound (~0.7 Hz). Thunderstorms are common sources of local infrasonics, especially if the waves remain on the line-of-sight transmission mode. According to Goerke and Woodward (1966), severe weather storms, including thunderstorms, can generate from 0.01 Hz to 0.05 Hz waves of from 1 dyne/cm^2 to 4 dynes/cm^2 for several hundreds of kilometers, especially in the direction of the weather movement. They conclude that infrasonic waves of this type originate from the leading edge of the storm, probably in the region of maximum convective forces.

Man-made Sources of Infrasound

Any weather-related infrasonic effect could be influenced by the variety of man-made infrasound sources already in the vicinity. Sensations of nausea and irritability were associated more than 40 years ago with 1 Hz vibrations from diesel engines and large propeller shaft movements. Infrasound is also known to be generated from defective ventilator shafts, air conditioners, powerful sirens, buses, bridges carrying heavy traffic, turbojet engines, and large rocket cones.

Perhaps the most common and persistent infrasonic source, especially for frequencies between 1 Hz and the hearing range (about 40 Hz), is the moving automobile. Infrasonic noise within a car with closed windows can reach peaks of from 110 db to 120 db at 16 Hz, especially on freeways; louder values occur with windows barely opened. The highest intensity band usually ranges between 2 Hz and 16 Hz (Stephens, 1969).

Biobehavioral Relevance

The weather-organismic relevance of infrasound is coupled with its long distance propagation and organismic penetrability.

These waves have the potential to induce weather-related symptoms that occur hours before changes in traditional meteorological parameters: wind speed, barometric pressure, temperature, and humidity. Although infrasonic amplitudes are very low in comparison with levels required at those frequencies to induce hearing sensations, their resonance patterns are compatible with human whole-body mechanical vibration. This potential for resonance interaction (the imparting of mechanical energy from the wave to the human body) and its great penetrability place infrasound and other extremely low frequency phenomena in an interesting theoretical position.

Correlational Studies

Feelings of sickness, nausea, and dizziness have been blamed upon infrasonic vibrations generated by stormy weather in several cities (Brown, 1973). The complaints appear significantly more frequently among people living in highrise apartments or working in tall singular structures with heights approaching the wavelength (or some harmonic) of infrasound. A simple calculation indicates that buildings between 34.4 meters and 344 meters tall would be especially prone to these problems during windstorms. Since wind velocity increases with the height above the ground, some tall narrow structures could be prone to a constant level of background infrasound.

Direct correlational data between infrasound intensity or frequency and human behavior has been limited, mostly because of the relatively few infrasound measurement stations. Using data from the Chicago, Illinois region, Green and Dunn (1968) found correlations of +0.37 and +0.49 between automobile accidents and days of moderate and intense infrasonic disturbances, respectively. Similar correlations of +0.33 and +0.50 occurred between moderate and high infrasonic intensities and the rate of absenteeism among school children.

Correlational Criticisms

The study by Green and Dunn, which is similar to other less rigorous reports, is compatible with the assumption that increased infrasound induces general undesirable behaviors. The relationship between the behavioral measures and infrasonic intensities is both positive and intensity dependent (increasing with intensity level).

A primary limit of these studies has been the lack of lag-time correlations between the infrasound levels and behavioral changes. If infrasound is indeed a primary variable for the display of these behaviors, a systematic temporal relationship should exist. Preferably, the correlation should be strongest before the major and obvious changes in weather occur. If the onset of the behavioral changes and the infrasound are coincident with other weather changes, the problem of confounding variables becomes paramount. In this situation, cloudiness, wind speed, and barometric pressure cannot be excluded.

Experimental Studies

Earlier studies by Gavreau, Condat, and Saul (1966) indicated that human exposure to from 1 Hz to 10 Hz infrasound was associated with the quick onset of fatigue, dizziness, irritation, and nausea. With a special infrasonic whistle designed by these French scientists, oscillation amplitudes became so large that visceral vibrations during exposure led to serious medical problems.

Human subjects exposed to infrasound generated by a special infrasonic speaker at frequencies between 1 Hz and 20 Hz at from 115 db to 120 db (100 dynes/cm^2) intensities displayed 30 percent to 40 percent slower simple reaction times. Frequencies between 5 Hz and 60 Hz at 104 db generated reports of nausea, apprehension, and slight dizziness in some subjects. Short duration exposures to comparable frequencies at intensities up to 150 db (about 5,000 dynes/cm^2) generated by a rocket engine, evoked painful vibrations in the chest wall, gag sensations, blurred vision, and respiratory alterations.

According to Stephens (1969), the distinction between airborne infrasound and whole-body vibration from local mechanical sources becomes less important as the frequency drops to below 100 Hz. Psychophysiological responses are very similar. Infrasonic vibrations from wall oscillations at 10 Hz at 115 db elicited dizziness in about 5 minutes. Around 10 Hz, other experimenters have reported an odd psychological response in exposed subjects, not unlike slight intoxication. The subjects reported a paradoxical lethargy and euphoria during which time judgment was impaired.

Such frequencies are often generated within a car traveling at around 100 km/hr. During driving tests, subjects exposed to these frequencies reported that they knew they were speeding but that they just didn't want to stop. Some scientists have suggested that added infrasound from transient weather-related sources could contribute to "infrasonic stupor," thus increasing the probability of traffic accidents (or any accident highly coupled with maintained concentration).

Problems with Experimental Studies

The burst of infrasound studies has clearly demonstrated that weather-related symptoms can be generated by exposures to infrasound of intensities higher than normal. Major methodological limitations have been the failure to use blind procedures and the lack of systematic intensity and frequency manipulations.

Most of the studies have involved acute exposures. Experiments must still be conducted to test the effects of long-term exposure to lower, more natural intensity infrasound. Until precise data are available concerning the ranges of infrasonic frequencies associated with particular types of weather changes (low pressure masses, frontal agitations, and so on), generalizations from experimental studies completed thus far will be limited.

Habituation times to infrasonic bathing are still to be found as well. Do the symptoms associated with experimental infrasound exposure decrease over time? Does the normal person habituate to their presence? This problem is important in light of the normal tendency for weather-related complaints to occur during meteorological changes and to decrease with time. Presumably, infrasound generated from an approaching unstable air mass would inpinge upon human populations for from hours to days.

Possible Mechanism of Detection

The mechanism by which infrasound might influence the human being is not precise. Certainly, the ear may mediate some parts of the infrasonic signal. Inner ear complications were apparently quite typical in submarine personnel exposed to the 1 Hz infrasound caused by suction strokes of engine cylinders. However, the possibility of whole-body absorption of these waves becomes attractive when one realizes that the human body vibrates normally within the infrasonic range.

Measurements of body vibration indicate normal oscillations between 6 Hz and 20 Hz with amplitudes of about 5 micrometers (μm) when the muscles are relaxed and of about 50 μm when the muscles are tense. Hyperthyroidism, fever, and related thermogenic processes increase the normal frequency of these rhythms. After death, these vibrations occur for about an hour. In general, a direct relationship exists between muscle contraction, body temperature, and the frequency/amplitude of these vibrations.

Applied infrasounds from the environment would be readily absorbed by the human body if they approached the vibration pattern of the human body. The maximum disparity between weather and body signals that can be tolerated before the potential effect becomes minimal is not yet known. Statistically, however, one would expect some individual's body vibrations to overlap with the infrasonic fields associated with some weather changes. Whether the intensities of the natural signals can be biologically effective, even when summated over the body surface and over several hours, must also be determined by experiment. Apparently, some marine creatures respond to microvibrations produced by low pressure areas over the Atlantic. According to Tromp (1963), infrasonic vibrations around 1 Hz with displacements of from 1 μm to 20 μm are associated with increased fish activity as measured by increased catch rates.

8

ELECTROMAGNETIC STIMULI OF THE WEATHER MATRIX: ELECTRIC FIELDS AND ATMOSPHERIC IONS

Electromagnetic stimuli within the weather matrix can be viewed as electric, magnetic, and electromagnetic phenomena. The differentiating factor between these three arbitrary divisions is primarily time. When either a magnetic field or an electric field is varying quickly in earth time, the two forms are difficult to distinguish. Time-varying or wave-propagating electric fields are associated with the induction of magnetic fields, and changing magnetic fields are associated with the induction of electric fields.

The principal form of electric phenomena is the charge, the unit for which is the coulomb. Charges are displayed as two discrete forms: positive and negative. Charges exert a force over distance through their accompanying electric fields. When large concentrations of charges are separated by space containing fewer charges, potential differences or voltages result. When the field generated within these spaces is quite intense, single charges (such as air ions) can be influenced significantly. Current is defined as the amount of charge moving across or within some specified surface per unit time. This flow of charge is measured in coulombs/second, which is equal to one ampere (A).

ELECTROSTATIC FIELDS

Suppose, in the laboratory situation, you have two metallic plates parallel to each other and separated by one meter. You attach the cathode of a battery to one plate and the anode of that battery to the second plate. The current rushes to both conductors (plates), but then can go no further to complete the circuit since air is not a good conductor. The charge remains on both plates and produces a potential difference or voltage between them.

Such disparity of charge and, hence, potential difference exists on a gigantic scale between the earth's surface and the very conductive upper atmosphere. Between the earth's (negative) surface and upper atmosphere (positive), a potential difference is generated, called the atmospheric electric field. The mean (vertical) value of the field strength near the ground varies markedly as a function of both universal and local factors. According to Israel (1973), it is a function of the specific resistance of the air at the measurement point, the columnar vertical resistance of the air, and the potential difference between the atmospheric electric equalization layer and the ground.

Variations in Electric Field Values

Mean values for electric field strength usually range from 120 volts per meter (V/m) to about 150 V/m as one ascends vertically. At two meters above the ground, the total potential would be 240 V. At heights attained by an aircraft, the potential difference between it and the ground can reach 100,000 V. The actual variation at ground level is subject to a large number of periodic and aperiodic phenomena superimposed upon each other.

Diurnal variations can display single peak and trough forms or double peak forms. The difference between the height and trough of the field changes is from about 40 percent to 60 percent of the total value. In central Europe, for example, the minimum clear weather field value is from about 100 V/m to 110 V/m between 0200 and 0400 hours local time and between 140 V/m and 150 V/m between 1700 and 1900 hours local time.

All atmospheric electric elements display a very pronounced annual variation, consisting of a single fluctuation with winter maximums (December to January) and summer minimums (July to August). The difference between the highest and lowest values is about 40 percent in the northern hemisphere. Secondary and statistical 27-day (solar rotation) and 11-year periodicities have also been reported in atmospheric electric field strengths.

Weather-related Variations

Since the columnar resistance of the air suspended or moving between the two boundaries of the earth-upper atmosphere capacitor contributes to total-field strength level, it is not surprising that differential air masses influence this phenomenon. If the stability of the vertical profile of an air mass were very low, significant fluctuations in the atmospheric field values or enhancement of variations already present would occur.

Most weather is associated with earth-ionospheric polarities typical of fair conditions. However, during light rain the potential gradient generally reverses polarity (direction) for protracted periods of time. Such reversals may occur during warm fronts, occlusions, or stratified cloud cover without discontinuities and are a sign of absent or minor uniform vertical movement of air (Israel, 1973). Light snowfall does not produce such reversals.

Strong and rapid fluctuations of field direction occur during cloudbursts combined with cold fronts, thunderstorms, and other forms of strong, rapidly varying vertical movements. During thunderstorms, the intensity of the gradient can achieve values exceeding 100,000 V/m. The passage of a front or the local exchange of air masses without precipitation can generate a singular rise and fall of the potential gradient (several hundred V/m within an hour.

A variety of man-made and local structures influence atmospheric field potentials and direction. Any structure jutting above a flat surface, such as trees, buildings, or even people, will distort the field potential. Lines of equal potential, for example, 120 V/m lines, will not pass through these objects, but will bend around or over them. Metaphorically speaking, the electropotential lines bend around an object, especially conductors, like water moving around a boat (the water does not move through the boat).

House walls, indeed any contained structure with relative conductive abilities greater than air, act as a shield against the changes of atmospheric fields. According to analytical studies, less than a microvolt (mV) per meter (10^{-6} V/m) would enter the average room from an atmospheric gradient that is 100 V/m outside. Even assuming a linear relationship, the outside potential must reach 10,000 V/m or more to increase the inside values by 1 mV/m. Slow, time-varying fields (less than 1 Hz) of 100 V/m would induce inside fields of about 10^{-3} V/m.

Biometeorological Relevance

The direct effect of atmospheric electric fields, in addition to electrocution from lightning strikes, is still a matter of controversy. Electrostatic fields, even slowly fluctuating varieties, are suspect in biological situations for a number of reasons. First, the induction of any current within the living system (the presumed mechanism of effect) is much too small, according to many theorists. Secondly, the electric fields of the atmosphere near ground level are subject to extreme distortions around the biological semiconductor. The most effective biorelevant property of electrostatic fields appears to be their determinant influence upon the direction and velocity of air ions.

Potentially, however, electric fields in the atmosphere contain sufficient weather-associated, specific variations to be considered— even if mechanisms are not apparent at this time. Since the direction and magnitude of these fields are so weather-dependent, they should be considered as possible conditioned or combination stimuli, much as the barometric pressure–relative humidity effect. By themselves, atmospheric potentials may be relatively ineffective. But in combination with another factor in the weather matrix they may be bioeffective.

Correlational Data

Electrostatic fields generated just before thunderstorms have been correlated with the curdling of milk or related colloids, inactivity of insects, irritability/restlessness of school children, enhanced traffic/industrial accident rates, and blood clotting times. No doubt, the confounding factors of contiguous increased relative humidity, decreased barometric pressure (usually), and elevated temperatures must be considered.

The accumulation of charge and the increased atmospheric potential around ground level before thunderstorms and snow squalls can reach significant levels as mentioned. Since air is not very conductive, very large potentials must occur before charge flows through or across an air gap. The probability of the flow is increased when charge can collect disproportionately, such as around points or sharp edges. Tops of steeples, some buildings, and even tips of leaves can reach near-discharge potentials. People in these situations report tingling sensations, hair standing on end, and odd feelings, including those of impending doom.

Experimental Studies

Few experimental studies have been concerned with electrostatic field values comparable to those generated by weather conditions. Fields in the order of 3 million V/m are known to kill bacteria, most likely by influencing the cell membrane. Fields of about 10,000 V/m have been reported to enhance silk production in silkworms. The possibility that 100 V/m electrostatic fields turned on and off every 12 hours can act as zeitgebers has been claimed by some experimenters (Dowse & Palmer, 1969).

Direct effect of atmospheric-level electrostatic fields upon the brain has had popular appeal. It seems to be common sense that a field of 1,000 V/m could influence the measley 100 millivolts generated by the nerve cell. However, the error lies in the space dimension. Although the potential difference between the inside and outside of the nerve cell membrane averages 100 millivolts, this charge disparity is distributed across a space only 10^{-8} m wide. If the same charge density were increased to a meter width, the net potential difference would be about 10 million V/m. At these intensities, the atmosphere field of 1,000 V/m becomes measley.

Steady Potentials of the Human Body

Since the billions of neurons in the human brain display a more or less steady-state condition (resting membrane potential) for long periods relative to 1 millisecond activity periods, one would expect a display of this condition over larger spaces. Whereas the resting potential of a single neuron may extend within less than a micrometer, a steady field should be formed over centimeter areas around millions of neurons. An analogy is the electric field induced around a bundle of several thousand separate strands of insulated wire.

Steady potentials of several millivolts exist across distances of a centimeter or more of the human brain. Isopotential lines of varying intensity forming a focus (analogous to closed isobars forming a cyclonic or anticyclonic condition on the weather map), emerge in brain parts prone to epileptic discharges. High gradients appear transiently during the consolidation of newly learned information.

However, d.c. or steady potentials exist between the nervous system and parts of the body. Potential differences of from about 10 mV to 30 mV exist between the backbone (spinal cord) and the distal parts of the hands and legs. According to Becker (1965, 1969, 1972), the spinal cord is positive with respect to the hands and legs, while the front of the head is negative with respect to the back of the head. Other voltage differences taken by Ravitz (1962) with slightly different instrumentation demonstrated potentials of between -30 mV and +30 mV between the human head and chest. The polarity of the difference (whether the head were positive or negative with respect to the chest) was quite variable and reflected gross behavioral changes.

Experimentally induced alterations in these potentials elicit several weather-related responses. A decrease in the potential difference between the front and back of the head is associated with less concentration and more mistakes on simple reaction-time tests (Friedman & Taub, 1969). Comparable shifts occur during sleep and depression. Unfortunately, they may be linked to the likelihood of suggestibility or responsiveness to hypnotic instructions.

Ravitz (1962), in studies involving much less control against experimenter bias and equipment artifact, reported steady-potential shifts in human psychiatric patients as a function of lunar phase. According to his measurements, reported in both graphical and numerical forms, d.c. variations of from -60 mV to +60 mV potentials reflected lunar phase. During new moon and full moon, the otherwise low-level potential difference between head and chest would display exaggerated magnitudes.

It is interesting that direct application of very small direct currents elicits behaviors similar to those associated with normal steady-potential fluctuations. As reported in a review paper by R. A. Herin (1968), applications in the order of from 10 mV to 100 mV or from 0.1 mA to 1 mA across the head or between the head and arm are sometimes, but not always, associated with dreamy states, gradual loss in consciousness, or mood alterations.

Whether the steady-potential patterns over the human body reflect meteorological changes must still be demonstrated experimentally. Animal studies certainly point in this direction. Comparable potential differences have been found across the bodies of many living systems, including trees, insects, and salamanders. Behaviors from these species appear to be coupled with environmental electrical fields.

Some theorists have suggested that the d.c. or steady state potentials of the whole body are a primitive communication network that mediates slow, dull pain of the arthritic variety. Entire nerve trunks across the length of the arm and legs, rather than specific pain receptors, are the likely candidates of mediation. Only at the more intense levels of stimulation do unpleasant experiences result. At stimulation levels below awareness, significant information about environmental change could still be processed. The person may respond to these changes by feelings, hunches, or anticipations of meteorological events to come or in progress some distance away. Pairing of these cues with the later presentation of the stimulus, a fundamental learning process, would be a necessary condition.

Induced Atmospheric Electric Field Currents

When a person is within a building, the walls, which have a comparatively lower resistance than the air, act as conductors and shield the room against the static atmospheric electric field. Changes in this field are ineffective, according to present knowledge.

Suppose the person is outside, directly exposed to the field gradient. From a physical perspective, the person is an electrical conductor. Since she forms an elevation (from about 1.5 meters to

2.0 meters), the equipotential surfaces of the atmospheric field bend around her as they do with trees and other obstructions. The equipotential surfaces do not pass through the person (Israel, 1973). In fact, since the skin is quite conductive, any induced effect would be not within the person but along a very thin shell of the outer skin.

Calculations of current induced from moving through the electric field (the presumed mechanisms of biological effect) vary from 10^{-12} amps (A) to 10^{-10} A. From conventional understanding of electromagnetic induction effects in living systems, these values are too small.

The work of Robert Becker (1972) and his colleagues suggests that low-level current induction may be more important than suspected. Some tissues respond to currents of about 1 millimicroamp (nanoamps) or about 10^{-9} A. Higher or lower currents are less effective or destructive. Whether such currents could be induced within deep tissue or whether extremely peripheral (outer skin) inductions could influence deeper tissue is still questionable. However, their results clearly demonstrate one factor: the assumption of a linear relationship between current intensity and biological effect must be based firmly upon data, not theory.

Arguments for and against the feasibility of very low-current (less than one microamp) effects upon living systems have been made since the turn of the century. The most well known exponent of this effect, H. S. Burr (1972), claimed that the growth of embryos could be influenced by these fields. Still other experimenters have claimed that axon growth can be stimulated or modified by application of similar d.c. currents.

Theoretical and psychological barriers by twentieth century scientists have prevented a general enthusiasm for Burr's contentions. For a phenomenon to be considered real, it must be easily replicated, even by the most clumsy scientist. When every positive study seems to have been balanced by one negative study, interest wanes. Without replication by many members of the scientific community, any phenomenon remains in a continuous state of controversy. Plus one minus one is never an impressive tally.

SMALL AIR IONS: PARTICULATE ELECTRIC FIELDS

Certainly, the gaseous component of the atmosphere is the most persistent stimulus bathing the boundaries of the human being. Although the vast majority of the approximately 10^{19} molecules per cc of air maintain an electric balance, a few molecules, between 10/cc and 10^4/cc of air, show electric disparity. Molecules that have gained or lost charge (electrons) are called ions.

In an absolute sense, the numbers of ions per cc of air are trivial. After all, they are diluted by a factor of at least fifteen orders of magnitude. The thick boundaries of the connective tissue (skin) could easily absorb and neutralize atmospheric ions with minimal biological consequences. However, recent studies indicate that the labile lung lining, the often forgotten human boundary which processes in the order of 10,000 liters of air per day, may respond to air ions.

Types of Ions

As Hans Dolezalek (1977) aptly states, "Ion is not equal to ion." Ions are distinguished by several different criteria depending upon instrumentation. From the criterion of charge, ions are either positive or negative. Positive ions contain fewer electrons than they would in a balanced molecular state. Negative ions contain more electrons than they would in balanced molecular state. In general, the mobility of negative ions is greater than that of positive ions by a factor of 100:1. According to Krueger and Reed (1976), each small ion possesses a unit charge of about 10^{-19} coulombs and, under optimal conditions, has a life span of several minutes.

Ions are often differentiated according to size and/or speed. Small ions are defined as charged particles that move about 1 cm/sec in an electric field of 1 V/cm, while larger ions, such as Langevin ions or large hydroscopic particles, are less mobile. They migrate at speeds around 0.001 cm/sec within a similar intensity field. Large, slow ions may contain more than a single unit charge, a factor that confounds some measurement procedures.

The chemical configuration and status of ions are extremely complex and variable. Configurations can change within the order of milliseconds and are determined by the variable types of molecules in the vicinity. Although water molecules are a common species that aggregates, from numbers of from one to eight, around ions within about a microsecond, other chemicals can contribute as well. In a normal atmosphere, the ions consist of $H_3O^+O_2^-$ (or some variant) and CO_4^- (Krueger, Andriese, & Kotaka, 1968). Organic systems add more complex ions to the milieu. However, this variable chemical nature of ions is seldom considered in biometeorological studies.

Concentrations and Sources

Clean country or mountain air contains from about 2,000 ions/cc to 3,000 ions/cc and rarely exceeds 10^4/cc (Krueger & Reed, 1976).

Small air ions combine with atmospheric pollutants to form La ions, resulting in depletion of small ions. Consequently, smal. concentrations around industrial centers can drop below 100/cc. These values are only about 10 percent of the concentrations foun in residential districts.

Increased ionization of air can be produced by friction generated between ice or snow particles, between falling water droplets (near waterfalls), or through windstorms. Whereas electric-tension discharge, such as lightning, creates about 50 times more negative than positive ions, air brushing past a metal surface through a pipe produces more positive ions. Ultraviolet light produces more negative ions, especially ozone, while the cracking of petroleum gases releases positive ions.

Special winds supply increased ion concentrations. Anticyclonic conditions, such as hurricanes, characterized by humid damp air are more prone to generate negative than positive ions. On the other hand, hot dry winds blowing over land, especially dry desert soils, are copious sources of positive ions. The Foehn, Sirroco, and sharav are classic sources.

The human being is both a direct and indirect source of air ions. Frictional interactions of shoes across carpet or of clothes against themselves or the skin generate several thousands of volts for short periods across very small distances (less than a mm). The accumulated charge can dissipate into the air or can be removed relatively quickly as low current when the person touches a conductor, such as a doorknob. Evaporation of water and organic molecules into the air from the skin is a common source of positive ions.

Maczynski, Tyczka, Marecki, and Gora (1971) noted that heavy positive ions tend to accumulate within business offices during working hours. In addition to the not surprising increase in carbon dioxide concentration (from 0.06 percent to 0.09 percent), large positive ions increased from about 500/cc to about 2,000/cc, although variability was quite large. Small positive ions increased from 6/cc to an average 27/cc, while negative ions increased from 42/cc to 58/cc; these changes were insignificant statistically. These researchers concluded that positive ion concentration increased and ionic hygenic index became worse as a function of the duration of human activity. The increase in large ions was dependent upon the amount of dust pollution in the room air.

Biometeorological Relevance

The sudden availability or depletion of positive or negative ions to or from the lung lining is more than a general step function.

Inhalation of air ions is like the sudden injection of a chemical into the body volume. Considering the area of the lung lining, the more than 10,000 liters of air breathed daily, and the passage of all the body's blood through the lungs every 30 seconds, absorption of any bioeffective chemical through the respiratory system should have significant consequences.

Air ions are chemical as well as electrical stimuli. They act directly at specific chemical partitions of the organism, within the complex configuration of molecules that maintain both structure and function of the body. Since air ions are unbalanced electrically, they are potent sources of electron exchange. Depending upon polarity, atmospheric ions can behave as oxidizing agents (chemicals that remove electrons from body chemical reactions) or reducing agents (chemicals that add electrons).

Unlike many other nontemperature stimuli of the weather matrix, feasible mechanisms of bioion interactions have been postulated. Since the reactions would occur primarily at chemical levels of discourse and the human tissue is approximately 80 percent water, most models assume the typical reactions of aqueous solutions. Krueger suggests that positive ions form complexes with available hydrogen ions, such as $H^+(H_2O)$ or H_3O^+, while negative ions form complexes with available hydrogen or oxygen ions, such as O_2^- (H_2O) or $OH^-(H_2O)$. If these chemicals were formed for short periods in body fluids, significant local distortions in pH (a measure of available H^+ or OH^- ions) could occur. Without buffers, additional H^+ ions would push the pH of the fluid toward greater acidity, while additional OH^- would increase the pH toward alkaline values.

Correlational Studies

A predominance of positive ions or a depletion of negative ions, a frequent condition of hot special winds, has been correlated with the serotonin irritation syndrome. This syndrome is composed of the following symptoms: migraine, nausea, irritability, vomiting, amblyopia, hyperperistalis, edema, conjunctivitis, and congestion of the respiratory tract.

Weather-sensitive patients who display these symptoms during periods of accelerated positive ion accumulation in the air, even from 24 hours to 48 hours before other weather changes, display the serotonin syndrome. According to Sulman, Levy, Lunkan, Pfeifer, and Tal (1978; also see Rim, 1977), the behaviors are reminiscent of altered serotonin activity in the midbrain. Those patients excrete abnormally large amounts of serotonin in the urine during positive ion periods.

Sulman and his colleagues successfully treated the disorder, in this case associated with the sharav, by administering serotonin-blocking agents. Inhalation of large concentrations of negative ions also appeared to reduce the syndrome. Controls for suggestibility were not reported.

Similar symptoms have been reported by populations dependent upon radical heating to maintain living temperatures during severe winters. Cold dry air from the outside becomes relatively more water absorbent (drier) when heated within living quarters. When the air is blown through metal pipes and ducts, copious numbers of positive ions can be generated. Without humidity controls, very dry, forced-blown air, even at comfortable temperatures (for example, from 65° F to 70° F) can become a quick source of positive ions. During nocturnal hours, the effects are usually increased due to both increased heating because of a drop in outside temperature and static inhalation by the sleeping person.

In general, correlational studies concerning negative ions report opposite effects. Ozone may occur in small concentrations (10^3/cc) following thunderstorms and electrical displays. Although reports of euphoria and well-being are paired with these conditions, several other factors, such as cooler temperatures and fewer dust particles in the air, confound these observations. Huntington (1959) insisted that feelings of well-being, enhanced activity, and consequently increased economic-social productivity were accompanied by influxes of ozone.

Experimental Studies

The vast majority of experimental studies concerning positive and negative ion effects have suffered technological and methodological flaws. Many of the earlier studies involved ion generators that produced toxic materials or uncontrolled numbers and sizes of ions. Classic biometeorological parameters, especially air flow, humidity, and temperature, were not controlled. Only recently have ion studies involved some type of control against expectancy effects and protection against experimenter bias.

The general belief in ion research is that positive ions are associated with nasty effects while negative ions are associated with good effects. In general, at concentrations within normal ranges, this statement is probably correct. However, at higher ranges, typical of oxidant levels during severe smog conditions, negative ions, such as ozone, can severely damage respiratory tissue, indeed any exposed tissue.

Early studies in which human subjects were maintained within a closed space for several hours reflect the pattern of ion results. Concentrations of positive ions between 10^3/cc and 10^6/cc, from a variety of unspecified sources, have been associated with complaints of itching nose and/or eyes, pharyngitis, dryness of mouth, and difficulty breathing. Influxes of negative ions at comparable concentrations relieved the symptoms. "Good" sensations and euphoria have been reported. It is not clear whether the negative ions produced the effect themselves or whether the removal of positive ions produced a contrast enhancement of any small mood change. Not all subjects reported the ion effects.

Knoll, Eichmeier, and Shon (1964) exposed volunteers to several concentrations of positive and negative ions. Subjects exposed to 2,000 ions, positive or negative, displayed reductions in simple reaction time. Higher concentrations (10^6/cc) of positive or negative ions were not associated with systematic changes in reaction times. The main effect was increased variability of response latencies, that is, more frequent extreme decreases or increases in reaction time.

Individual variability, as have all nontemperature related biometeorological ailments, has been a critical variable in ion studies. Silverman and Kornblueh (1957) found that subjects exposed for 30 minutes to air enriched with 1,800 positive or negative ions/cc showed decreased alpha frequencies. Whereas some subjects demonstrated the decrease with negative ions, some showed the effect during positive ion inhalation, while still others showed the effect with both. Inhalation of larger concentrations (10^6/cc) for 5 to 10 minutes was correlated with an unpredictable increase or decrease (increased variability) in alpha frequency according to Eichmeier (Knoll, Eichmeier, and Shom, 1964). The alpha frequency effects were quite small and can be simulated by simple relaxation or by closing the eyes.

Animal studies have been more controlled and appear to give more perspective to the intensity of the effect. At concentrations typical of weather changes, negative ions in the order of 1,000/cc reduce emotional behaviors and facilitate simple maze learning. The effects, however, are very weak and are comparable to third-order variables (factors that control marginally significant effects) in animal studies.

Bachman, McDonald, and Lorenz (1966) suggest a possible reason for the conflicting literature. They measured activity in rats as a function of ion concentrations that ranged between 15 ions/cc and 10,000 ions/cc. Whereas activity increased with concentrations of up to 1,000 ions/cc, this behavioral measure decreased at higher concentrations.

A nonlinear or U-shaped relationship between the concentration of any chemical and behavior is a general pattern in pharmacology.

Whereas small or optimal dosages of a chemical may stimulate behavior to some accepted criterion, higher concentrations of the same drug are toxic. The crux of the ion problem has been: What is the effective concentration and what is the toxic dosage? The danger of the problem is: What is the difference in concentrations between the effective and toxic levels?

The Everpresent Expectancy Effect

People unfamiliar with the powers of verbal instruction and social cuing upon human behavior and physiology always underestimate the role of these factors in human experiments. In situations that are unclear and/or heavily loaded with emotional anticipation, such as expectancy of relief from a malady, the verbal instructions and social procedures of the laboratory situation can influence significantly what the person thinks, feels, and experiences. Such experiences seem real.

The problem of expectancy and social structure in laboratory experiments involving people is especially relevant to atmospheric ion studies. Physical scientists, engineers, and "leisure scientists" have comprised the bulk of bioion investigators. Since they are not generally familiar with psychosocial variables, this problem cannot be overemphasized. The following experiment was designed by this author to test the significance of expectancy and social cues upon mood rating.

Expectancy was generated through vocal instructions while social cuing was mediated through stooge behavior (a stooge is a member of the experimental team who is familiar with the purpose of the experiment). A total of 40 female subjects were exposed to one of two expectancy conditions, positive (E+) or negative (E-), and to one of two social behavior conditions, positive (S+) or negative (S-). Consequently, there were four groups of ten subjects each.

The subject and the female stooge, who was a part of the experiment, were brought into a standard test room in the Department of Psychology at the University of Wisconsin—Madison. The room was bright, well-ordered, and contained three chairs and a table. An "ion device" with bright lights and impressive looking gadgetry was displayed on the table. However, the "ion device" was a sham; it did not generate ions of any form.

Each subject was then read either a positive (E+) or negative (E-) description of what other experimenters had found in ion studies. In the positive expectancy profile, the subjects were told that atmospheric ions enhanced thinking and produced feelings of well-being, and, in some cases, even euphoria. In the negative expectancy profile, the subjects were told that ions made people feel depressed, sad, and sometimes ill. Both expectancy conditions were concluded with an emotional jag instruction: the subjects were told the study was tremendously important because ion concentrations were associated with world wars and a peak in solar activity and ion concentrations was approaching. A fictional graph was shown to indicate the relationship between ion concentrations and the occurrence of war.

After the instructions, the experimenter plugged in the dummy ion apparatus and left the room for 20 minutes. Within about five minutes, the stooge began one of two prearranged acts. In the positive social behavior act (S+), she began to shoot rubber bands across the room and to act "light-headed." Over the next ten minutes, she made airplanes, walked around the room, and giggled, mentioning how good she felt.

In the negative social behavior act (S-), the stooge merely sat in place for the first few minutes, then received permission to sketch the subject and began drawing on a artist's tablet. After about five minutes, she began to look sad and talk about how badly she felt. With about five minutes left in the segment, she would tear up the drawing of the subject and run out of the room crying.

A number of different mood-rating tests were administered at the end of the 20-minute session. Subjects were asked to check words from a long list of descriptors to describe how they felt. Positive words included overjoyed, energetic, clear-thinking, playful, vigorous, lively, active, and elated. Negative words included sluggish, annoyed, sad, washed out, tired, grouchy, angry, and blue. Twenty cartoons were rated on a scale from 1 (not funny) to 5 (very funny).

Two general predictions were made. First, in light of the work by Schacter and Singer (for example, 1962), one would expect the negative expectancy to generate more arousal (uneasiness) than the positive expectancy. Since arousal is a primary condition for the cognitive (psychosocial) aspects of the situation to influence mood, the greatest discrepancy in mood should have been between the negative stooge and positive stooge conditions for the negative expectancy only. Secondly, since humor is determined heavily by expectancy and by the arousal boost of instructions, one would expect humor ratings to reflect the direction of the expectancy more than the social behavior of the stooge.

The results confirmed both predictions. With any score less than 12.0 indicating a negative mood, the ten subjects of the positive stooge condition with negative expectancy averaged 14.4 ± 2.5 ($\pm$ values are standard deviations), while the subjects from the negative stooge condition with negative expectancy averaged 9.1 ± 2.0. The other two groups S+E+ and S-E+ averaged 12.7 ± 3.0 and 12.2 ± 2.5, respectively. For the cartoon ratings, the E+S+, E+S-, E-S+, and

E-S- groups averaged 2.9 ± 0.5, 3.0 ± 0.4, 2.3 ± 0.3, and 2.2 ± 0.4, respectively.

Although the results are statistically significant, the important feature of the experiment is its implication. Report of private experiences, such as mood or humor, are influenced by the instructional content and social behaviors displayed in the experimental situation. In this experiment, merely listening to a brief description about the effects of atmospheric ions and/or being exposed to behaviors that are presumably reflective of ion effects produced significant alterations in behaviors classically associated with bioion studies.

The Krueger Studies

A. P. Krueger, E. J. Reed, and their colleagues have been reporting various effects of air ions upon living systems for more than 20 years. The core of the controversy is expressed by these authors in a recent review paper (1976). They have worked with several tissue and animal models.

For people who prefer parity, Krueger's study with rabbit trachea cilia and different polarity ions is exemplary. Mucous flow in this tissue was measured to decrease from about 3 mm/minute to 1 mm/minute when positive ions were applied (10^4/cc), but to increase (5 mm/minute) when negative ions were applied in similar concentrations. The tracheal mucous flow associated with the positive ions could be duplicated by injections of serotonin. Serotonin effects were reversed by applications of negative ions or by reserpine, a known inhibitor of serotonin activity.

Similar results were found in blood levels of serotonin (Krueger et al., 1968). Mice exposed to 10^5 positive ions/cc displayed serotonin elevations from control values of about 5.6 μg/cc of blood to about 6.0 μg/cc. This marginal but statistically significant elevation was evident within 24 hours of exposure and did not increase further during six days of exposure. Negative ions exposures (10^5/cc) were associated with lowering of serotonin blood levels to about 5.0 μg/cc within the first 24 hours. However, the values returned to normal within two days.

Brain levels of serotonin did not reflect the polarity preference. In another study (Krueger & Kotaka, 1969), groups of mice were exposed to one of six conditions: one of two types of ions—negative or positive—and one of three concentrations—10^3/cc, 10^4/cc, or 10^5/cc. Within 12 hours of exposure to 10^5 ions/cc, brain serotonin levels dropped from usual control values (from 0.95 μg/gm to about 0.85 μg/gm). The decrease occurred for both negative and positive ions. The decrease in brain serotonin levels were apparent for all but one concentration of the six conditions after 72 hours of exposure, not after 24 and 48 hours of exposure.

Although the blood and brain serotonin levels are comparable with absolute values reported in other studies (Barchas & Usdin, 1973), the effects are very, very small. The maximum change in brain serotonin associated with 10^5 ions/cc in Krueger's study is similar to the normal range in daily (circadian) fluctuation (Rechtschaffen, Lovell, Freedman, Whitehead, and Aldrich, 1973). Compared with the administration of well known serotonin-interacting (inhibiting) drugs, such as parachlorphenylalanine, which reduces serotonin from average levels of 0.9 μg/g to 0.5 μg/g of tissue, the ion effect is trivial.

Krueger's pursuit of the bioion effect has recruited many different experimenters and techniques. One example (Krueger, Kotaka, & Reed, 1971) was his measurement of influenza death in mice following positive ion exposures. Continuing a study by the controversial Tchijevsky, published in the 1930s, he found that mice injected into the nose with influenze virus died more quickly when they breathed air containing 10^5 small positive ions/cc.

According to Krueger, small and positive ions may mediate their effects by penetrating and impinging upon the upper portions of the respiratory tract. No doubt, common complaints from alleged ion sources are congestion in the nose, pharynx, and, with prolonged exposure, bronchi. However, a more convincing support of this contention is the ineffectiveness of both positive and negative ions on the pathological progress of influenza viruses when the upper respiratory tract was bypassed.

Integrating the research of ion effects, a general pattern appears to have emerged. Unipolar high concentrations (greater than from 500/cc to 1,000/cc) of positive ions or ion-depleted air contribute to respiratory diseases and to the serotonin syndrome. Air pollution from industrial activities contributes to ion depletion by adding nuclei to which the small ions, negative and positive, can attach to form larger Langevin ions. Negative ions (from 100/cc to about 1,000/cc) are associated with improvement of these problems.

The Serotonin Hypothesis

Excitement about the correlations between atmospheric ions and blood/brain serotonin levels, even if the relationship is not systematic, is warranted. If atmospheric ions do exert direct or even secondary changes upon brain serotonin content, their total and long-term impact cannot be ignored. Serotonin, also called 5-hydroxytryptamine, is a major chemical transmitter candidate in the brain. A vast wealth of literature exists demonstrating clear and undisputed relationships between brain serotonin levels and aggression, sleep duration, pain sensitivity, depression, headaches, sexual activity, psychotic behaviors, and general autonomic function.

Special anatomical studies have shown that much of brain serotonin originates from a local concentration of cells in the brain stem. This cluster of cells, the median raphe, sends axons throughout the entire brain. Serotonin or the chemical from which it is formed appears to be transported along the axons to the terminal areas. When stimulation of the neuron occurs due to electrical stimulation from other neurons or from changes in blood chemistry, serotonin is released from the terminal areas. In this manner, neuronal information from the brain stem is passed to other brain structures.

The distribution of serotonergic neurons throughout brain space is not homogeneous. Relatively larger concentrations of serotonin nerve endings are found in the hippocampus, the septum and amygdala, the hypothalamus, and, more moderately, in the cerebral cortexes. Hippocampal activity is correlated especially with the onset of dreaming during sleep or the inhibition of dreaming during the waking state. Hypothalamic activity integrates the vast number of vegetative functions, from temperature control to water balance.

Stimulation of the amygdala and/or septum, two of the major structures of the limbic system, evokes many overt and covert emotional behaviors in the human. Aggressive outbursts, sexual activity, and bouts of depression are well known correlates of these structures. Some theories of emotion indicate that the amygdala is associated with feelings of self-preservation and "meaning," while the septum is associated with frank sexual fantasies. Stimulation of these structures in unrestrained human subjects can elicit sudden feelings of cosmic meaningfulness for the most trivial event or sudden feelings of sexual desire.

Disruption of serotonin content through direct interference with the stem (median raphe) nuclei and/or by modification of the limbic system structures would be expected to produce alterations in behaviors associated with these structures. If serotonin functions as the chemical mediator of the central parasympathetic nervous system in the brain proper, as suggested by Brodie and Shore (1957), suppression of serotonin synthesis or levels should result in a compensatory overcontrol by the central sympathetic nervous system. In general, the results appear to agree with this hypothesis.

The most common experimental means to reduce brain levels of serotonin in rats and people has been the administration of parachlorophenalalanine (PCPA). Animals given large concentrations of PCPA demonstrate hypersensitivity to pain stimuli, more frequent aggression (as measured by mouse killing), greater likelihood of convulsions, and more emotional behaviors. The effects of PCPA can be reduced somewhat by high dosages of caffeine.

The most persistent and obvious behavioral disruptions following PCPA-induced depletions of brain serotonin levels are enhanced sexual behaviors and insomnia. Male rats demonstrated increased frequency of all classes of sexual behavior: mounting, erection, ejaculation, and pelvic thrusting. These hypersexual behaviors, like many of the PCPA-induced behaviors, were abruptly reversed by administration of 5-hydroxytryptophan, the chemical from which serotonin is made. Prolonged EEG activation and sleep disturbances are manifested in both reductions of dream and deep-sleep periods. That insomnia is coupled with irritable aggression in human subjects is well documented.

Depressions of serotonin levels in human brain have been correlated many times with psychotic and self-destructive behaviors. Analysis of people who committed suicide demonstrated lower levels of serotonin and other biogenic amines, such as norepinephrine, in the brain stem relative to controls. Schizophrenic patients also demonstrate lower serotonin brain stem levels. However, any psychopathology with a predominant symptom of depression, a primary precondition for suicide, demonstrates lower levels of brain stem serotonin.

Volunteers who have taken PCPA have reported severe headaches, anxiety, irritability, and depression. They became more paranoid, suspicious of attendants, and generally distrustful of social interactions. Severity of hallucinations induced by the hallucinogen LSD were found to increase when PCPA had been preadministered.

Sleep and dream disruption have been documented in a number of serotonin-related disturbances. During the typical sleep sequence, a normal person displays successive patterns of electroencephalographic patterns. Rapid eye movement (REM) sleep is associated with dreaming and is commonly displayed during Stage I sleep patterns. The electrical activity of the cerebral cortex during REM is very similar to that of waking or conscious behaviors. Deeper stages of sleep are characterized by very low EEG activity. States III and IV are commonly called slow-wave sleep since they are associated with delta waves (between 1 Hz and 4 Hz). The average person progresses through each of these stages from every 90 minutes to 2 hours or from about 3 to 5 times per sleep period.

A large portion of schizophrenic patients have little or no slow-wave sleep. Comparable types of insomnia or sleep disruption occur following administration of PCPA. Unfortunately, other conditions prevent or decrease slow-wave sleep. Asthma, pregnancy, depression, mental retardation, and increased age also reduce or eliminate slow-wave sleep. Although serotonin has been implied in these conditions, it has not been demonstrated clearly.

Brain stem serotonin depletion or decreased activity allows dreams to occur during other "states of consciousness." Schizophrenic patients are known to display more REM activity while awake, a condition that could generate very rich images indistinguishable from reality. PCPA administration releases REM or dream experiences into waking behaviors.

The relevance of the ion-serotonin hypothesis to biometeorology should be clear by now. Depression, irritability, sleep disturbances, headaches, suicides, greater response to pain, and related behaviors are historical and crosscultural correlates of weather conditions known to generate bursts of atmospheric ion concentrations. Add the relatively well established observation that serotonin levels may evoke hyperthermia/hypothemia in response to cold/hot temperatures, and biometeorology is given a discrete handle on a diffuse subject.

One could rationalize a variety of observations. For example, schizophrenic patients, already characterized by low serotonin levels, should show greater behavioral instability to weather changes that load their environment with ions. Depressed patients, with low serotonin levels, should be pushed over the statistical edge to suicide with a little added depletion from ion levels. Older people should show greater weather sensitivity. S and V (Sex and Violence) should protrude from the statistics. The rationalization could go on and on.

However, the serotonin hypothesis is not the answer; it is only a well asked question. In short, serotonin in the brain stem appears to be synthesized in the following manner. Tryptophan obtained through food circulates through the blood bound to the serum fraction. Tryptophan then enters the brain across the blood brain barrier, where the amount determines the total serotonin levels. Tryptophan is synthesized into 5-hydroxytryptophan by the enzyme tryptophan hydroxylase; PCPA depletes serotonin by inhibiting this enzyme. In turn, 5-hydroxytryptophan is synthesized into serotonin by another enzyme, a decarboxylase. Serotonin is then quickly broken down by a monoamine oxidase after it is released from the nerve terminals.

Serotonin levels can be altered by influencing the monoamine oxidase activity, synthesizing enzyme activity, or various chemical pools along the pathway. Although this may explain partially why various assays are inconsistent in different bioion studies, the variable sight of ion effect along the pathway makes isolation of mechanism very difficult. If the effect were mediated by cyclic changes in enzyme activity and only secondarily reflected in serotonin levels, the effect would be more difficult to isolate.

More fundamentally, is the serotonin effect just a secondary phenomenon? Are there other brain chemicals that respond more specifically to ions and to other parts of the weather matrix? For example, the enzyme that synthesizes serotonin may also synthesize

dopamine, a precursor to norepinephrine in the brain. Norepinephrine neurons cluster in a brain stem area called the locus ceruleus.
This area is well known as the controller of dream onset as well as
a variety of other behaviors relevant to biometeorology.

There is a great distance between a general statement such as,
"Atmospheric ions affect brain serotonin," and the actual demonstration of the relationship. There is a very big difference between a
general sell phrase such as, "People commit suicide when the special
winds blow," and the pharmacological prevention of this unfortunate
consequence. The difference between general semantics and specific
description is the difference between blaming the weather and doing
something about it.

9

ELECTROMAGNETIC STIMULI OF THE WEATHER MATRIX: MAGNETIC FIELDS AND EXTREMELY LOW FREQUENCY ELECTROMAGNETIC FIELDS

In situations in which time variations are minimal as defined by known equations, magnetic phenomena can be distinguished clearly from what have been called electric phenomena. Magnetic fields have much greater penetrability than do electric fields, and they influence different classes of materials.

Several units of magnetic intensity or magnetic flux density have been used. One gauss = 1 oersted = 10^{-4} webers/m^2 = 100,000 gamma = 10^{-4} Teslas (T). The weather matrix contains two general forms of magnetic phenomena: those with variations greater than one second and those with variations less than one second. The first type is primarily of geomagnetic origin and includes geomagnetic fluctuations or magnetic waves generated by very large geological processes. Variations of the second type, generally between 1 Hz and 100 kHz, are correlated with classical weather processes, such as convective alterations and air mass movements.

Electric currents are induced within appropriate objects when they move across magnetic flux lines or when magnetic flux lines move across them. The intensity of the induced current is a function of the intensity and direction of the applied magnetic field, the frequency of the (electro)magnetic field, and the conductive capacity (as measured by resistance or resistivity, and so on) of the material. Different organ systems display marked variations in conductivity and penetrability by environmental magnetic fields.

GEOMAGNETIC FLUCTUATIONS

The earth can be considered a large dipole, with magnetic flux lines (lines of force) leaving one pole and entering another. Since the total force of the magnetic field in space is related to the number of flux lines per unit volume, the intensity of the magnetic field is higher near the poles, where the lines emerge, than near the equator, where they diverge. The major source of the geomagnetic field is assumed to originate from complex dynamo processes at depths approaching the earth's core.

Actual values for average geomagnetic intensities range from about 25,000 gamma near the equator to about 70,000 gamma near the poles. There are significant local variations due to man-made constructions and natural depositions of ore. Such magnetic anomalies can reach 300,000 gamma (3 gauss) or drop (shield) the local field to near zero values. Although subject to secular variation over periods of decades, these values can be considered the steady-state or baseline levels upon which short-term variations are superimposed.

Traditionally, the geomagnetic field has been discussed in terms of total force, declination, and inclination. The total force is described in three-dimensional space. Primary distinction is made between the vertical and horizontal components with a secondary division of the latter component into north-south and east-west. Declination is the angle in degrees that a horizontal compass needle deviates from geographical north. Inclination is the amount of dip in the compass. At the magnetic north pole, the compass dips straight down (inclination of 90°), while at the magnetic south pole, the same needle points straight up. A variety of maps have been drawn to demonstrate areas with equal inclination angles (isoclincs), declination angles (isogonics), or directional forces. Samples are usually published in most elementary geophysics or geomagnetism books.

The Magnetosphere

The magnetic fields of the earth extend far out into space. If the geomagnetic field assumed a perfect dipole geometry, one would expect the extension to be about the same on both the sun and dark (antisolar) sides of the earth. However, the geomagnetic geometry is modified by constant "pressure" from the solar wind.

The solar wind is the metaphorical name given to the constant flow of plasma, primarily fully ionized hydrogen, that results from the continual expansion of the sun's corona into the solar system. On solar quiet days, the velocity of this wind is about 400 km/sec with a density of from about 1 atom/cc to 10 atoms/cc at the distance of the earth's orbit. On solar active days, the velocity increases to 1,000 km/sec with densities as high as 30 atoms/cc for periods in the order of 10 hours.

Figure 16 depicts the shape of the geomagnetic field as it might look (if we could directly see magnetic flux lines) several hundred thousand miles away. Due to the "pressure" exerted by the solar wind, the geomagnetic field is compressed on the sun side so that the boundary (the magnetopause) between the earth's field and the solar wind is about 10 earth radiuses wide. The net effect of this compression is to increase the intensity of the geomagnetic field from about 10 gamma to 20 gamma more than a nonwind condition.

Whereas the geomagnetic lines are compressed on the sun side, they are extended far into space on the night side. As shown in Figure 16, the inertial and coercive forces of the solar wind pull the anti-solar side of the geomagnetic field away. This geomagnetic tail has been measured to extend to distances of from 20 to 40 earth radiuses, perhaps approaching the lunar orbit. Charged particles from the solar wind can be trapped for long periods within the neutral sheet created by the disproportionate extension of north and south polarity flux lines.

Geomagnetic Storms

A geomagnetic storm is a period in which the components of the earth's magnetic field display exaggerated, irregular deviations from the normal quiet day pattern. The deviations display periods lasting from several seconds to hours, while the total storm duration ranges from a day to weeks. A sample magnetogram from a quiet to a storm condition is shown in Figure 17.

During storm conditions, the horizontal component can vary between 10 gamma and 1,000 gamma (about 0.02 percent to 2 percent of the total field intensity) and the declination can vary several degrees around baseline values. Since magnetograms are only pictorial representations of geomagnetic displays, numerical "equivalents" are derived for calculation use. The K index, which is a semi-log scale based upon the severity of the most disturbed geomagnetic component, is a common measure. Other indexes are A and C. The subscript "p," such as in K_p, A_p, C_p refers to a global measure. For the most part, they are all highly intercorrelated. They are published monthly in the Journal of Geophysics and by central data groups.

Geomagnetic storms are divided into three components: the sudden commencement and initial phase, the main phase, and the recovery phase. The initial phase is associated with a plasma cloud, which can be considered a compression of the quiet day solar wind due to a flare or other source of energetic disparity interacting with

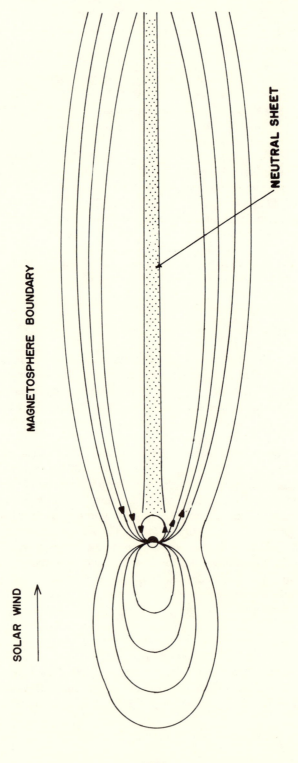

SOLAR WIND →

MAGNETOSPHERE BOUNDARY

NEUTRAL SHEET

Figure 16 Idealized conception of the earth's magnetic field in outerspace, demonstrating compression on the solar side and distention on the antisolar side.

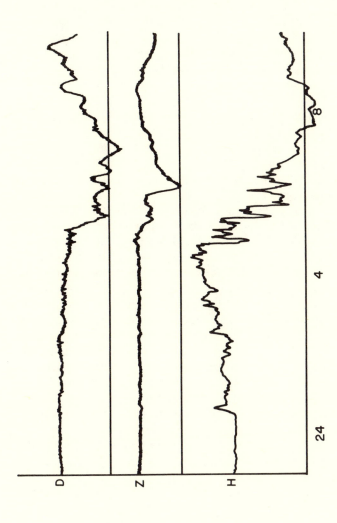

Figure 17   A sample magnetogram demonstrating the change from a quiet (before 24 hours) to a storm condition. The letter designations are D = declination, Z = vertical intensity, and H = horizontal intensity. A vertical displacement of 1 mm for either of the latter two measures is equal to about 3 gamma. The absolute Z and H intensities for this record are 52,687 gamma (0.53 gauss) and 19,569 gamma (0.20 gauss), respectively. The X axis indicates time in hours.

the magnetosphere. On the ground, the net effect of the greater compression is an enhanced field intensity of from 10 gamma to 100 gamma.

The initial sudden increase in compression is mediated to the earth's surface to increase the horizontal component by an average of from 20 gamma to 30 gamma over a few minutes. In some instances, this sudden commencement may be characterized by horizontal component increases of from 50 gamma to 100 gamma. Continuation of the solar wind disturbance maintains the increased compression of the field, creating the initial phase of the storm, which lasts from about two hours to eight hours. Some storms may not be preceded by discrete sudden commencements but, instead, show slower increases.

The main phase of the storm is characterized by a decrease of the mean value of the horizontal component to below the normal prestorm value. An average main phase decrease is from about 50 gamma to 100 gamma, although excursions of over 1,000 gamma can occur for short periods. Excursions of several hundred gamma with periods of 0.5 hours are typical and, in some instances, may approach 1 Hz. The main phase of the storm lasts from about 12 hours to 24 hours. Like the initial phase, the main phase is noisy, characterized by irregular fluctuations in horizontal and declination components.

Several theories indicate that the decreased intensity on the surface of the earth's magnetic field is due to the outward distension of flux lines into space. This extension has been considered a consequence of pressure from the hot plasma freshly trapped in the field or of outer flux lines shorn away by the passing plasma cloud. As a result, the flux lines close to the surface move outward, decreasing the average intensity. The rapid fluctuations of the field seem to be caused in part by the varying pressure of solar wind turbulence. Very short fluctuations are evoked by waves propagated from the outer magnetosphere to the surface within about 100 seconds.

The recovery phase occurs after the plasma cloud or enhanced solar wind disparity has passed the earth. It consists of a slow recovery or relaxation of the horizontal component to prestorm values. The recovery period is associated with the cooling of the plasma in the geomagnetic field. Characteristic times of recovery are about one day, although periods up to a month have been required following intense prolonged storms.

Periodicities in Geomagnetic Activity

Periodicities in geomagnetic activity occur during solar quiet and active days. During solar quiet periods, small diurnal fluctuations with average amplitudes of a few gamma occur due to the effects of lunar tidal forces in the upper atmosphere or to solar heating. The latter periodicities are related to ionospheric convective currents passing through the geomagnetic field, inducing small electromagnetic forces on the earth surface. Such variations are lost in the noise during solar active periods.

Geomagnetic activity displays an interesting annual pattern, even during quiet day or solar quiet year intervals. Two maximums in activity are apparent during January and June, while two minimums occur during March and October. The average peak to peak (interval between the highest monthly values and the lowest monthly values) amplitude of this variation is about 20 percent of the mean level (Russell & McPherron, 1973).

The most well known and strongest periodicities in geomagnetic activity are coupled with sudden (that is, from a few minutes to hours) disturbances in the solar wind. Marginal but significant increases in geomagnetic activity occur when the moon passes between the sun and earth (new moon) and during full moon. Stronger storms are correlated with the passage of sectors from the solar magnetic field across the magnetosphere. Other perturbations occur every 27 or 28 days and appear to be coupled with solar rotation.

Certainly the vast majority of the variance in geomagnetic activity is a consequence of solar storm conditions. These periods, characterized by massive discharges of radio waves and energetic particles, displays of flares and plaques, and increased sunspot size and numbers, display periodicities of from 11 to 12 years. Geomagnetic disturbance profiles display similar periods.

Geomagnetic Activity Correlates with Weather

Recent studies clearly indicate reliable correlations between the severity of geomagnetic disturbance and traditional weather parameters. Using data from the British Isles, Beynon and Winstanley (1969) found that periods of increased wind velocity (gale speeds) followed increased geomagnetic activity by one or two days, while decreased wind speeds followed geomagnetic quiet periods by a similar period. Between 20 percent and 30 percent decreases below mean values in wind speed occurred after days with about half the average geomagnetic activity values. A similar relationship was reported as early as 1888 by B. Stewart and W. L. Carpenter.

A much larger spatial effect of geomagnetic disturbance upon developing air masses has been shown several times by Roberts and Olson (1973). They noted that wintertime low pressure troughs moving over or formed within the North Pacific region between two and four days after a bright aurora or geomagnetically disturbed day showed trends toward abnormal pressure gradients. Troughs formed during these periods demonstrated vorticity areas between 50 percent and twice as large as those formed during quiet days. Although the physical mechanism is not clear in detail, these results clearly link the early formation of potential weather disturbances with geomagnetic storms.

In a series of correlational and theoretical studies based upon data from Japanese weather stations, K. Takahashi (1966) found a persistent but complex trend between geomagnetic disturbances and precipitation. Decreases in precipitation were found on geomagnetically disturbed days or about two or three days after an intense solar flare in most of the stations sampled. The effect was complicated by seasonal effects. Not all stations showed the effect, which was about 10 percent around the mean precipitation values.

Assuming that the decreased precipitation on geomagnetic active days was a consequence of greater precipitation on previous days, Takahashi (1966, 1968) performed key day analysis on ten weather stations, including Denver, Colorado; Siska (Soviet Union), and Helsinki, Finland. With key day analysis, specific types of days are set equal to zero regardless of when they occurred. Days before them are given values of -1, -2, or -3, and so on, while those occurring afterward are given values of +1, +2, +3, and so on. In this case, key days were all days of strong geomagnetic disturbances. Takahashi reported that most of the stations showed increases in precipitation between 10 percent and 30 percent of average values, from one to three days before geomagnetic disturbances. Since strong solar flares precede strong geomagnetic disturbances by two or three days, the suggestion is made that solar disturbance may somehow contribute to precipitation over areas with cloud conditions prone to this possibility.

The next logical step in the geophysical analysis of frontogenesis and precipitation profiles will be to integrate the approaches and to ask specific questions. To what degree do season and latitude contribute to the total variability of the geomagnetic-solar effect? Does repeated or persistent solar activity, such as during sunspot maximums, increase the probability of labile air masses as well as increased precipitation? Is the effect additive or nonlinear when combined? Can such small magnitude phenomena when repeated frequently over several years alter the baseline in local weather? Will the areas (in the order of $10,000$ km^2) prone to meteorological extremes, now assumed to be chance displays, remain the same or will these loci change geographical position?

Biometeorological Relevance

Unlike atmospheric electric fields, geomagnetic fields can penetrate living systems in the open environment as well as in most dwellings. Disturbances in these background fields, assuming frequencies of from 0.01 Hz to 0.1 Hz, fluctuation intensities of 100 gamma, and typical conductive values for the human being, could induce currents certainly as high as Becker values deep within the human organism.

Geomagnetic activity must be considered a part of the weather matrix not only because of its direct effect upon the human system, but also due to its recent correlations with weather. When weather changes are associated with geomagnetic disturbances a few days before, any weather-associated disorder must be evaluated carefully. Biobehavioral changes induced or evoked by geomagnetic disturbances from two to three days before could be misconstrued as results of the weather on the day of change.

Correlational Studies

In the order of 100 studies have been published since 1900 concerning correlations between geomagnetic activity (storms) and various forms of human and nonhuman animal behavior. This author has critically evaluated these studies (Persinger, 1974). Analyses involving human samples have involved daily fluctuations over short periods of a few months or yearly fluctuations over decades or centuries. The latter studies, although frequently quoted by popular press, contain gross multicontributory behaviors, such as social unrest and wars. In most cases, geomagnetic disturbances have been inferred from sunspot numbers rather than correlated quantitatively.

Researchers in the Soviet Union have reported many correlations between health and geomagnetic activity. For example, significant positive correlations have been found over monthly and yearly epochs of analysis between geomagnetic indexes and over 2,000 heart failures for one city (Novikova, Gnevyshev, Tokareva, Ol, & Panov, 1968). The peak to peak component overlapping with geomagnetic activity is from about 2 percent to 20 percent of the mean heart failure. Unfortunately, the baseline level of heart failures was subject to large fluctuations over a time when many political and nutritional factors were changing. In another study, 1,000 observations of acute glaucoma demonstrated a linear increase with deviations of the horizontal magnetic field component. Translated into approximate correlation values, these reports would display coefficients of about 0.3.

Unfortunately, the minor enthusiasm generated by geomagnetic disturbance-disease correlations is, for the most part, canceled by studies that utilized very large samples from many geographical areas. The majority of negative results have been reported by large U.S. health organizations with access to large data banks. Such procedures allow data selection without the bias risks of personal data collection. For example, a recent study by Lipa, Sturrock, and Rogot (1976) found absolutely no relationship between geomagnetic disturbance (as measured by the A_p index) and coronary disease and stroke in either Washington, D.C., Phoenix, Honolulu, or San Francisco. The analysis involved mortality for these diseases between 1962 and 1966. Unlike less rigorous studies, these experimenters normalized the mortality data by removing weekly, seasonal, and secular variations, working on the assumption that any true geomagnetic-mortality correlation would be a short-term phenomenon.

The possibility that geomagnetic activity may be correlated with general increases in human activity, normal or abnormal, has been suggested by Friedman, Becker, and Bachman (1963). They correlated a total of 28,642 psychiatric admissions to seven hospitals over more than a 4-year period with geomagnetic activity measures. Since the time between the onset of psychiatric disturbance and hospitalization varied, analyses were completed according to 7-, 14-, 21-, 28-, and 35-day periods. Geomagnetic activity of these time periods was correlated with admission rates during the same time periods at the different hospitals.

Their results, considering the use of such a crude measure, are compelling. For 7-day analysis intervals, the correlations between psychiatric admissions and geomagnetic activity were not significant. However, as the analysis interval increased, the correlation coefficients increased as well. Whereas the 14-day interval correlations ranged from +0.07 to +0.22 (median = +0.12), the 28-day intervals correlations ranged from +0.22 to +0.35 (median value = +0.28). The values are statistically significant. Although low, they are consistent.

Criticisms of Human Correlational Studies

Two additional problems are encountered in the correlational studies published to date. First, the data are usually presented in a biased manner, in order to emphasize the parallel between geomagnetic fluctuation and the behavioral measure. A classic example of this procedure was published in 1935 by Traute and Duell. They found that the number of nervous disorders increased markedly from about one to three days after a geomagnetic disturbance. Figure 18A is a reproduction of the manner in which they presented the data. Figure 18B adds the rest of the dependent variable axis and gives a more realistic view of the geomagnetic activity contribution. Note that the apparently large effect in Figure A becomes a slight ripple when presented properly.

The second major criticism of the human correlation studies involves the time interval chosen for analysis. The problem is two-edged. When the analysis interval is too small (daily magnetic activity indexes with daily behavioral measures), small but general relationships can be lost. Friedman, Becker, and Bachman's data demonstrate this hazard. On the other hand, analysis intervals that are too great (yearly magnetic activity indexes correlated with yearly behavioral measures) increase the risk of including all kinds of confounding variables.

Such great analysis intervals require even greater total periods of study. While merely a year of data is required to detect most significant correlations with weekly analysis intervals (a total of 52 observations), at least 50 years should be involved when analysis intervals of a year are used. A single increase and decrease (one curve) in a behavior over a 12-year period with corresponding overlap in geomagnetic activity is hardly proof of relationship.

Animal Correlational Studies

Correlational analyses using nonhuman animals in laboratory situations are subject to more control against confounding factors, although, in general, the same criticisms apply. Fluctuations of about 1,000 gamma are correlated with alterations in the orientation of honey bees and other insects. Unfortunately, these alterations occur during periods of extreme geomagnetic activity as do many other factors, including radiation influx, ultraviolet light changes, and other stimuli known to influence insects. There is reliable evidence that diurnal geomagnetic changes may be one stimulus in the hierarchy of zeitgebers for invertebrates.

More quantitative measures from life systems, such as oxygen consumption, have been used to monitor geomagnetic fluctuations. One sample study was reported by Barnwell (1960). Over a 30-day period, daily oxygen consumption was measured in mud snails. Correlation analyses with geomagnetic indexes indicated significant relationships between this metabolic measure and geomagnetic activity one day (-0.54) and two days (-0.54) later.

Correlational analyses allow quantitative statements rather than general guesstimates. Sometimes, however, one may be faced with two distinct sets of data. Numerical analyses give some credence

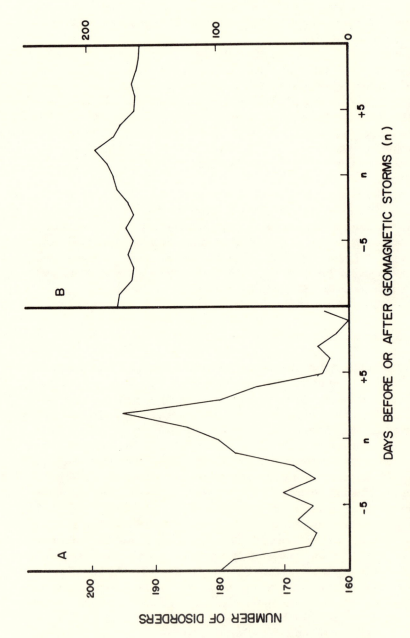

Figure 18   The effects of different presentations from relative versus absolute changes in human behavior as a function of time lapsed since intense magnetic storms day (day = 0).

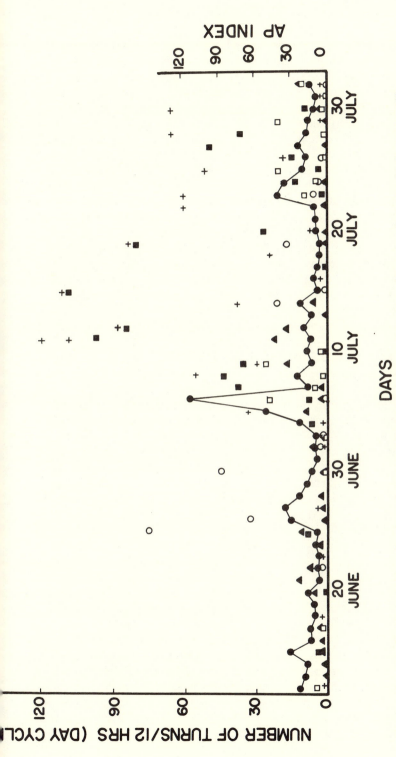

Figure 19   Daylight running-wheel activity in laboratory rats over several weeks during which an usually intense geomagnetic storm was registered (July 4 and 5, 1974).

Source: Modified from Persinger (1976).

239

to the conclusion that periods around severe geomagnetic storms should be treated as a population separate from periods of low or moderate activity. An example of this problem is shown in Figure 19. The Y-axis is day cycle of running activity in laboratory rats, a primarily nocturnal behavior. The Y-axis on the right is the activity of the geomagnetic field.

Whereas little light-period activity occurred before the geomagnetic storm of July 4 and 5, 1974, significant increases in activity occurred thereafter. The activity, which required about a week to peak, continued for at least two weeks after cessation of the geomagnetic event. Correlational analyses using daily intervals for activity measures before the storm and after the storm demonstrated no significant coefficients, even with lag analysis. Like the Friedman, Becker, and Bachman (1963) data, greater intervals would be used to elucidate the pattern. Optimally, with several years of this type of data, separate analysis should be made for severe storm periods (for example, weeks or months rather than days) and for nonsevere periods.

EXTREMELY LOW FREQUENCY ELECTROMAGNETIC FIELDS

Electromagnetic fields, waves, and impulses that occupy the frequency band between 3 Hz and 3 kHz have been called extremely low frequency (ELF). Very low frequency (VLF), from 3 kHz to 30 kHz, and ultra low frequency (ULF), less than 3 Hz, phenomena occupy adjacent wave bands. In this book, however, time-varying magnetic, electric, and electromagnetic phenomena ranging from 0.01 Hz to 100 Hz will be designated as ELF, since this range overlaps with major time-varying electromagnetic-biochemical processes in living organisms.

Characteristics

ELF fields may occur in nature as pure waves or may be presented as ELF pulses from VLF carrier waves. In the latter case, VLF waves, for example 10 kHz, are received as short pulses lasting from 10^{-1} to 10^{-4} seconds, with the number of pulses per second (pulse frequency) being in the ELF range. The sources of ELF and VLF fields are primarily ionospheric above from 4 Hz to 7 Hz and magnetospheric below those values. ELF phenomena associated with local weather conditions seem to span the entire ELF spectra. Artificial sources of ELF also contribute to electromagnetic (EM) noise.

Ionospheric origins seem to come mostly from lightning discharges called atmospherics or "sferics." The amplitude spectra of atmospherics show maximums from 30 Hz to 100 Hz and at 10 kHz with a minimum between 2 kHz and 3 kHz. At 5 Hz, the magnetic component is about 10^{-8} gauss (a milligamma), while the electrical component would be about 2 mV/m. By 400 Hz, these perturbations fall to less than 10^{-11} gauss and 2 μV/m, respectively. At the lower end of the ELF band at which the wavelength begins to approximate the circumference of the earth (40,000 km), a unique resonance system is produced. Power spectra of these "Schumann resonances" show maximums at 7.8, 14.1, 20.3, 26.4, and 32.5 Hz, with electric field strengths up to 1 mV/m. Power level decreases as the frequency increases, while the propagation parameters depend on the boundary characteristics of the ionosphere.

The principal amount of ELF energy remains within the spherical shell between the earth and ionosphere. Consequently, ELF waves in this mode can travel around the earth without appreciable attenuation. For example, a 1 mV/m wave at the source will decrease to about 0.9 mV/m after traversing a million meters within the spherical wave guide. The frequency of resonance depends on the distance between the source and receiver. ELF emissions are at times observed from near 4 Hz to 9 Hz. They are so termed because of the difference between these frequencies and the fundamental Schumann frequencies from 8 Hz to 14 Hz.

These resonances as well as VLF waves originate primarily from thunderstorm activities in the active areas of equatorial Africa, southeast Asia, and in a belt running from southern North America to central South America. Increases from 25 Hz to 100 Hz waves with from 10 kHz to 25 kHz components have been correlated with peak thunderstorm activity in the Amazon Valley and Central Africa. The major sources for at least some areas can apparently shift, depending upon season.

The intensity and number of ELF waves and pulses show diurnal and seasonal variations. Both 9 Hz fields in Germany and from 50 Hz to 300 Hz fields in North America displayed amplitude peaks between 14:00 to 16:00 hours local time and amplitude minimums between 00:00 and 08:00 hours. This variation was most pronounced during the summer months, during which time the peak magnitude was from 3 to 10 times that of the intensity of the minimums. Tromp (1970) has reported that pulses of from 3 kHz to 50 kHz waves of from 0.01 to 0.2 V/m exhibit a maximum between 23:00 and 0:700 hours with a minimum between 10:00 and 16:00 hours during January through March. A second maximum occurred during May through August at from 14:00 to 16:00 hours with a minimum of 0:800 hours. Tromp indicated that the number of impulses rapidly increased in May and June, with a maximum in August.

Reception frequency also seems to follow a day-night change. Surprisingly, König (1962), who summarized the work of several experimenters, found that fields below about 130 Hz were received more in the day while higher frequencies dominated at night. Attenuation for waves from 50 Hz to 300 Hz is about 2.3 decibels per megameter (db/Mm) in the day and 1.6 db/Mm at night, a value which depends strongly upon the local magnetic field dip angle. At night, west-east propagation shows less attenuation than east-west propagation.

ELF phenomena associated with geomagnetic disturbances are often sinusoidal wave forms that occupy the frequency range from 3.5 Hz to 0.15 Hz and are often called pearls, hydromagnetic (hm) emissions, or type 1 continuous pulsations (pc1). On the average, these events last about 35 minutes with amplitudes rarely exceeding a gamma (10^{-5} gauss). Waves in the outer magnetosphere often show higher intensities, of from three gamma to eight gamma. Diurnal variations are apparent, but no seasonal periodicities have been noticed. It has been suggested that geomagnetic micropulsations originate from hydromagnetic waves in the magnetosphere that propagate down the earth's flux lines to the surface.

Several authors have shown a correlation between weather conditions and ELF electromagnetic waves and pulses. König (1974) noticed that, unlike the 9 Hz waves (Type I) associated with strong lightning, waves from 2 Hz to 6 Hz (Type II) were top-waved and were connected with rain and heavy, deep-lying clouds. These frequencies did not show diurnal variation and were sometimes observed for hours with pulse intensities up to 0.1 V/m. Frequencies from 0.5 Hz to 2 Hz (Type III) were recorded with sinusoidal wave patterns, but their origin was unclear. Local lightning strokes were also recorded as impulses with about a half-second duration from trough to trough. During sunrise, a unique 9 Hz wave, which was later dominated by a from 3 Hz to 4 Hz band, occurred.

Reiter (1964) has published considerable data implicating the association between unstable air masses and pulses in the 4 kHz to 12 kHz and 10 kHz to 50 kHz spectra. In Europe, polar air masses were correlated with decreased numbers of atmospheric pulses in the summer and early autumn and increased pulses during winter and spring. On the other hand, maritime and tropical air masses decreased these pulse counts in winter and autumn and increased them in summer.

During stable weather conditions, ELF pulse frequencies of from 1 Hz to 3 Hz superimposed upon a 10 kHz carrier wave have been demonstrated. Local unstable weather conditions, such as the close passage of a cold or warm front, result in marked increases in the incidence of pulse frequencies from 30 Hz to 100 Hz superimposed upon a carrier between 10 kHz and 100 kHz. The latter fields varied from 0.01 V/m to 0.1 V/m and were propagated with retention of appreciable strength from 10 km to 100 km from the frontal or low pressure turbulence system. Ormenyi (1972) reported that the lower end of the ELF band (3 Hz) may show greater weather specificity, at least in Hungary. He found an increase in this band's power spectrum during cold front passages and labilization processes, but a decrease during warm front passages, upgliding processes, and influxes of subtropical air masses in the higher levels of the atmosphere. Wave patterns near the source area favor impulse cascades and statistical fluctuations in contrast to a nearly constant frequency series for very distant sources, in which case the lower ELF resonances are more pronounced. In some areas, the pulse frequency might vary from 100 Hz down to 0.01 Hz.

Local factors can also influence the frequency and intensity of ELF fields. These include local water level, mineral content, topography, and altitude. For example, gains as high as 60 db (for example, from 1 mV/m to 1 V/m) have been reported between water-poor valleys and higher plains with underground water sources. Changes of this magnitude have even been recorded between two points 100 meters apart, a marked local variation.

Biometeorological Relevance

ELF electromagnetic fields may be important biological stimuli because of their penetrability and long range of propagation. Compared with many other meteorogenic stimuli, ELF waves have a greater capacity to penetrate the microclimate of the average house. Faraday cages, which can significantly shield out radio frequencies, have minimal effect on the ELF band. According to the physicist Wolfgang Ludwig (1974), the transparency of brick houses to 10 kHz ELF pulse carriers is about 95 percent of 100 mV/m, while a large Faraday cage of 1-mm iron wire with 30-mm wide mesh is about 65 percent. An atomic bunker, on the other hand, constructed of 60 cm of concrete shows a transparency of less than 1 percent at these frequencies.

ELF pulse frequencies and waves occur in the wave bands of many electrical processes in mammals. Individual neural activity can occur between $\leq$ 100 Hz and 1,000 Hz, while electrical activity in large areas of the brain ranges from about 1 Hz to more than 50 Hz. In fact, König (1974) repeatedly mentions the striking similarity in wave form and frequency between ambient Type I (10 Hz) ELF waves and the human alpha wave pattern and Type II (from 2 Hz to 5 Hz) waves and delta wave patterns. Changes in EEG frequencies have long been associated with modifications of subjective behaviors, including consciousness.

The possible influence of alpha frequency electromagnetic fields on alpha rhythm-related behavior has been a major impetus for preliminary work in this area. Interaction options are attractive since slow potentials in the cortex can vary from less than 0.01 Hz to about 1 Hz. They have been implicated as important correlates for learning and memory storage. Even the microvibrations of the musculature in human organisms occur in this wave band, especially between 4 Hz and 12 Hz.

Correlational Studies

Reports of correlations between reaction time (RT) or behavior related to this measure, such as traffic accident rates, and the incidence of atmospheric ELF pulses or waves have received much popular attention. Early reports, primarily collected by Reinhold Reiter in the early 1950's and extended by H. L. König (1962) claimed that reaction times became quicker during the incidence of fields from 4 Hz to 6 Hz. These conclusions were based upon thousands of uncontrolled reaction times taken during a traffic show.

Close examination of these reports and similar claims usually exposes more than just the subjects. For example, in the reports by König, the changes in reaction time, even if they did involve thousands of subjects, are confounded by time of day and by unspecified local weather changes. The actual magnitude of change is very small. For example, the maximum decrease in reaction time over baseline during a maximum fivefold increase in ELF signal intensity was only 5 percent. The standard deviations of the RTs were not reported.

An improvement in correlational studies is characterized by the work of W. R. Ranscht-Froemsdorff and his colleagues. H. Brezowsky and Ranscht-Froemsdorff (1966), found that during periods of the year when the intensity maximums for 20 mV/m ELF pulses from 10 kHz to 100 kHz waves were present two times per day instead of once per day, the pattern of myocardial infarct became bimodal as well. A significant number of myocardial infarcts occurred before the minimums and shortly after the maximums on these days. The absolute magnitude of the effect was in the order of 10 percent. Possible confounding factors from unusual weather during these episodes were not determined.

R. Lotmar and Ranscht-Froemsdorff (1968), noting that respiration rates of rabbit tissue in test tubes showed considerable day to day variability, correlated tissue oxygen consumption values with intensity values for weather-related ELF signals. Intensities of

400 mV/m, impulses typical of labile weather conditions (low pressure areas), were significantly correlated (~ 0.7) with decreased respiration rates, while signals typical of fair weather (anticyclonic conditions) were correlated with increased respiration rates. It is not clear whether the latter effect was a compensatory response or a contrast enhancement, since the two signal types show great temporal proximity. Similar deficits in oxygen uptake have been noted during the passage of low pressure areas and maximums in from 3 kHz to 50 kHz 400 mV/m pulses within ELF frequencies.

Some of the most well-designed correlational studies have been completed by Wolfgang Ludwig and his associates (for example, Ehrmann, Leitner, Ludwig, Persinger, Sodtke, & Thomas, 1976). Assuming that pain reported by arthritic patients might be coupled with ambient ELF pulses from VLF carriers, Ludwig carefully correlated the relationship between subjective ratings of the patients and local atmospherics values. Although the two measures correlated significantly, the atmospheric values were also correlated with other more blatant weather changes, such as precipitation. In order to determine whether the ELF pulses were primarily involved with the subjective complaints, Ludwig constructed special covers containing light metallic materials under which the patients were requested to sleep. The materials were selected on the bases of his extensive work with electromagnetic shielding materials (Ludwig, 1973).

Unlike many other reports by experimenters less familiar with the difficulties of ELF magnetic fields and human research, Ludwig's results appear both empirically realistic and theoretically reasonable. He found that the shielding covers significantly reduced arthritic complaints. However, the effect was weak and subject to great individual variability. Some subjects, from a pool of patients with differential diagnoses, reported large improvement; others reported very little effect. However, no subjects reported an intensification of the complaints.

EXPERIMENTAL STUDIES WITH MAGNETIC FIELDS

In an area such as biometeorology, plagued by unclear diagnostic boundaries, multicontributory symptoms, and general insolubility, the behavioral scientist expects to find a normal human paralogic best described as "same treats the same." This form of magical thinking, through which the believer attempts to remove a mysterious disorder by applying a mysterious procedure, has dominated both serious and frivolous work with magnetic fields. Scientist and scoundrel alike have been attracted, perhaps by the semantic simplicity of the argument, to the thought that the invisible force of magnetism must be able to treat those ailments outside the discrete, particulate diagnoses of medical science. There is, after all, only a tenuous separation between the concept of an invisible electromagnetic force and spiritual power.

Many cultures contain weather remedies that involve application of magnetic fields to the location of irritation. These anecdotal cases remain firmly rooted in tradition since many of the diseases have inherent fluctuations of severity within from hours to days. Objective evaluation of the actual causative value of magnetic fields with such diseases is very difficult. The possibility that the disease/pain/irritation would have abated anyway, without the applications of magnets, is either not considered or rationalized within some popular concept.

Semiscientific literature abounds with reports of magnetic field effects. Review of these complicated and, for the most part, marginal experiments by so-called general scientists and nontechnical writers have cluttered the trade market with a mishmash of pulp and speculation. The consequences of exposing rats to static magnetic fields of 10,000 gauss (20,000 times the intensity of the earth's steady-state magnetic field) are compared casually with correlations between heart disease and local magnetic field intensity. Headaches and brain changes from energetic microwave fields pulsed within ELF ranges are nonchalantly pooled with the vague correlation of geomagnetic fluctuations. Invariably, the poorer or inappropriate studies are used to buttress a writer's private global theory displayed in a "magnetism can explain everything not yet explained" format.

The confusion is contained in the concept. The relationship between magnetic fields and biological systems is an idea that never matured. It is still primitive, diffuse, and devoid of the complex geometry of application that could change it from a superstition into a science. Whereas other primitive ideas, such as protopharmacology, became rational, practical phenomena through repeated quantification, isolation, and purification of effective stimuli, magnetobiology remained impotent. The critical determinants were never sought systematically, and the required technology was never pursued. At present, the total area of biomagnetism is analogous to early pharmacology, when biobehavioral effects were elicited by unpurified compounds collected from unspecified sources and given in unmeasured dosages through any mode of ingestion.

Singular rejection of magnetobiological effects or electromagnetobiological changes would be inappropriate as well. There are many experiments, involving high intensity (from 1,000 gauss to 10,000 gauss) magnetic fields, high intensity (from 1 kV/m to 100 kV/m) 50 Hz or 60 Hz electric fields, or ELF pulsed (amplitude modulated) microwave or radiowaves, that have shown clear, reliable, and, for the most part, deleterious changes in exposed biosystems. Within the conceptual dimension of EM interactions with biological systems, one must include the literally thousands of experiments concerned with ionizing radiation, especially X rays and ultraviolet rays. Reflexive rejection of electromagnetobiological effects would be throwing out the baby with the bath water.

However, if one selects experiments on the bases of field intensities comparable with or slightly higher than meteorological-geophysical values, field frequencies or signal patterns similar to natural conditions, and elementary methodological controls, the serious experimenter is left with about 100 published studies. If one further selects for studies with these criteria involving human subjects, the total number of experiments is reduced to less than ten.

Human Experiments

About half of all human experiments with weatherlike ELF electric or magnetic fields involve reaction time (RT) measurements. The general procedure requires the subjects to press a lever as quickly as possible whenever a stimulus, such as a light or tone, is suddenly presented. The interval between the onset of the stimulus and the depression of the lever is called the simple RT. During any one session, the RT stimulus can be presented periodically or un-expectedly, although the average subject quickly anticipates that the greater the elapsed time since the last stimulus, the more likely another stimulus will be presented.

Superimposed upon this simple testing procedure are ELF exposures. Some experimenters alter intervals of field-on and field-off, while others have altered frequencies of the field during succes-sive periods of exposure. In more well-controlled experiments, some subjects are exposed to sham conditions whereby all procedures are followed as usual (including turning knobs, if any), except the power cable to the field generator is disconnected. Electric field experiments require the subject to sit between plates or screen mesh across which the fields are generated, while magnetic field experi-ments require subjects to sit within or near coils of wire (Helmholtz coil or air core solenoid).

The simple RT experiment is attractive since the measurements are numerical and clear. Presumably, they share some relationship to the reaction time sequences displayed while driving a car or con-trolling a drill press. Quantitative support for this relationship is sparse. Simple RT measurements have a major limitation: very intense stimuli, environmental or psychological, are required to produce reliable changes in mean values. Depending upon the demand characteristics of the situation, an otherwise effective stimuli may have no effect. For example, if a macho male subject were told that the test were evaluating his masculine capabilities, no changes in RT may be apparent even though he feels odd or sick at the time of the field presentation. If the subject could not care less about testicular tenacity, he might display clear alterations.

That simple RT is an inappropriate measure of ELF effect is evident from the literature. Early experiments by König (1961) showed that presentation of from 3 Hz to 5 Hz fields around 4 V/m produced an average 30 milliseconds (msec) increase in RT, an apparently substantial effect. Considering that typical simple RTs are around 300 msec (0.30 sec) the absolute change is 10 percent at most. If a driver were traveling at 60 mph (100 km/hr), the delayed 30 msec would allow another meter of movement before braking began. Since braking distance involves tens of meters (several car lengths), the ELF effect is trivial. For comparison, alcohol consumption increases absolute simple RT by a factor of at least 3, from 300 msec to 900 msec. A car traveling at 60 mph would proceed another 18 meters during this time.

When more rigorous procedures are added to ELF human experiments, especially to control for the effects of just sitting and the implicit realization that something will happen about midway through the experimental period, the ELF RT effects are not seen easily. For example, Persinger, Lafrenière, and Mainprize (1974) exposed a total of 70 male and female subjects to alternating 10-minute sequences of either 3 Hz or 10 Hz electric fields generated between two metal plates. Two field intensities, 0.3 V/m and 3.0 V/m, and a sham field condition (same procedure with no electric field) were used. No significant differences in average (mean) reaction time were apparent between either frequency or any of the intensity or sham field conditions. In fact, 99 percent of the variance in the mean RTs could be explained by the U-shaped change in RT during the 40-minute session, individual differences, sex differences (a well-replicated phenomenon), or the pattern of light stimulus presentation, whether it was presented in predictable clusters with rest periods or distributed more evenly with no rest periods.

However, subjects exposed to the 10 Hz sequences displayed increased variability in RT (more frequent excessive RTs beyond the mean) than subjects exposed to the 0.3 V/m or sham field conditions. The effect was weak and accounted for about only 30 percent of the total variance of variability. Variability may indeed be the primary factor behind the failure to find ELF effects in RT experiments. Friedman, Becker, and Bachman (1967), who exposed more than 70 human subjects to either 0.1 Hz or 0.2 Hz magnetic fields between 5 and 12 gauss, generated by coils, also noted weak RT changes that could be enhanced when the data were normalized for individual differences.

Acute exposures in the order of a few minutes to ELF EM fields may not be sufficient to demonstrate their effects. Several German experimenters have found that human subjects exposed for days to weeks within situations to which ELF fields were periodically added or removed demonstrated reliable changes. Whereas long-term exposures may increase the probability of detecting any weak effect, they also increase the probability of including random, irrelevant, and infrequent confounding variables. The dilemma between detection or masking and acute or chronic treatments is not unique to biometeorology.

Even with this consideration, the experiments are compelling. Ranscht-Froemsdorff and Rinck (1972) exposed subjects for from two to four weeks within special climate chambers that could both shield external ELF pulses and produce them. In one of the first experiments of its type, they also added a second biometeorological variable: temperature. Subjects were exposed to either between 12 and 17°C, 17 and 22°C, or 22 and 27°C temperatures and periodically to experimentally generated "bad" weather impulses (from 10 to 100 kHz fields pulsed between 30/second and 100/second at 100 mV/m). Blood clotting time was the primary measure.

The effects were realistic and expected from the correlational data between ELF signals and blood clotting diseases, such as thrombosis, stroke, and hemorrhaging. The presentation of ELF signals contributed to the variability of blood clotting times. Temperature, strongly suspected from correlational data taken during Foehn conditions, interacted with the effects. Whereas subjects exposed to the temperatures between 17 and 22°C showed greater clotting variability with the field on than with the field off (null field), subjects exposed to the hottest ranges showed greater variability with the field off than with the field on; in the latter case, the field actually normalized the variability.

The second major pattern that substantiated previous correlational claims was the duration of the effect. In all cases, the effects either increased and decreased variability of clotting times, or were transient, peaking two to three days after the initiation of the ELF signal application and returning to normal or to preapplication levels shortly thereafter. These authors contend that their data explain why thrombosis-type problems seem to be potentiated during cold, unstable, and hence "bad" weather impulse-generating air masses, but are attenuated during warm, unstable air masses.

If the data were replicated and generalized, the differential interaction between ambient temperature and ELF impulses from unstable weather conditions could explain, even determine, the multitude of blood problems, from peripheral clotting to water imbalance, classically associated with barometric pressure changes. The problem of differentiation will be a tricky one, not only because of the close temporal proximity of barometric pressure and ELF signals, but also due to the powerful propensity for the human being to be conditioned.

The second major and most convincing series of experiments concerning ELF fields and weather-related behaviors is from the laboratories of Rutger Wever (1974). Although primarily interested in the intricate relationship between the daily or circadian rhythms of the different systems within the human body, Wever has tested the effectiveness of several zeitgebers or cues to which these rhythms may respond.

General activity, body temperature, and other physiological values of human beings in a normal environment show very clear 24-hour cycles. Activity is greater during morning and afternoon hours and decreases before sleep. Body temperature is highest, in general, from around noon to 6 p.m. and lowest between 2 a.m. and 4 a.m. Under free-running conditions (no cues), the circadian rhythms begin to average from about 24.5 to 24.7 hours. Consequently, lagging begins and the person may awake from 30 minutes to 45 minutes later each day. Without clocks or time cues, such as "good morning" and "good evening," he wakes up later and later according to external references, but "normally" according to his own internal cues.

Wever found that he could stop the internal desynchronization and prevent this lagging or, as an operant psychologist might say, bring the behavior under the control of another stimulus. When a 10 Hz square wave electric field of 2.5 V/m was presented continuously for 12 hours a day and removed for the remaining 12 hours within the isolation chamber, the subjects' activity and physiological rhythms began to follow the field within from one to three days. Instead of continuing the approximately 30-minute delay each day, the rhythms became synchronized with the onset and offset of the 10 Hz field. The data are clear and strong. Neither the application of a d.c. electric field of 300 V/m nor a magnetic field of 1.5 gauss could stimulate the powerful zeitgeber or resynchronization effects of the 10 Hz electromagnetic field.

Wever suggests that normal intensity variations of natural environmental 10 Hz fields (a thousand times less intense) also act as synchronizers or cues to the physiological rhythms of the human body. Transient interference or masking of these signals from bad weather ELF impulses or even from man-made sources, such as defective ELF pulse-generating equipment, could disrupt the synchrony of the body's rhythms. One would expect, from the isolation data, that the psychological consequences would be similar to jet lag (from crossing more than five time zones) with predominant symptoms of low energy, constipation, and sleep disturbances. People in self-contained, near-constant environments, especially invalid and geriatric populations, would be expected to be more prone to disturbances from weather. Since there is a clear tendency for physiological systems to respond to the most potent ambient zeitgeber, these people would have come under the control of and be responsive to disturbances in normal ELF variations.

The results of Ludwig, Ranscht-Froemsdorff, and Wever indicate a clear, if not a major, role of ELF fields in low-level meteorstatic human behavior. However, the effects of these fields are not immediate; they require time. Several days may be required before the effects of ELF fields can be detected. From the data collected in Germany, the time involved is from about one to three days, a characteristic period in weather changes.

Ludwig (1971) concludes that on the bases of the data available to date, intensive care patients should be placed in rooms shielded from both man-made and natural bad-weather ELF signals. He recommends ELF electromagnetic field production at 10 Hz during the day and at 5 Hz during the evening in hospitals, a suggestion commensurate with the studies of Wever. If the enthusiasm generated by the ELF research were reflected in strength of replication, we may one day find daily ELF signal reports as a part of normal community messages to warn susceptible patients.

An ominous possibility about which scientists shiver or from which they shun is synergism. There are borderline fears of unpredictable interactions and enhancement of effect when people are exposed simultaneously to natural ELF fields and the increasing amount of man-made electromagnetic noise pollution. W. R. Adey (1975) and his colleagues at the University of California-Los Angeles, for example, have found that 147 MHz (microwave) fields of only 1.0 milliwatt/cm^2, but pulsed at from 0.5 Hz to 30 Hz (amplitude-modulated), leave permanent changes in the EEG records of test animals from hours to days after the fields have been removed. How would natural fields interact with such effects? Synergisms between man-made EM noise and geophysical-meteorological ELF fields are a theoretical necessity; their practical demonstration will be an empirical nightmare.

Nonhuman Animal Studies

Human subjects are the optimal animal model for biometeorological research. Measurements of many traditional ailments asso-

ciated with weather require self-evaluative responses, such as mood
or pain. Whereas a behavioral scientist might legitimately compare,
with great reliability, the general monthly activity of rodents with
monthly general activity of human beings, comparisons between a
human feeling and a rat squealing are precarious. Nonhuman animals
studies are important in biometeorology since they can be conducted
under controlled conditions. In general, there are few quantitative
differences between the major physiological systems of monkeys,
humans, and rats.

E LF electromagnetic research with animals has demonstrated
the patterns and the problems of human data. In the last ten years,
Persinger, Lafrenière, and Ossenkopp (1974) have exposed hundreds
of rats during various stages of development to 0.5 Hz magnetic
field with intensity variations from 10 gamma (0.1 milligauss) to
10 million gamma (100 gauss). A multitude of behavioral tissue,
physiological, and blood measures have been recorded. Only the
general patterns will be discussed here.

Most of the behavioral tests have involved maintaining the animals within the experimental field or control conditions but testing
them outside the fields. The rationale for this procedure rested
upon concurrent interest in memory and the consolidation of memory
processes. In theory, training or testing the animals outside the
experimental field conditions while maintaining them continuously
in the field between testing allows determination of the role, if any,
of geomagnetic-like ELF fields on memory consolidation.

In several experiments, juvenile, young, and old rats were
exposed continuously from 10 days to 30 days to 0.5 Hz variation
intensities between one thousand gamma and one million gamma
(from 10 to 10,000 times natural intensities). On several test days,
they were removed from the nagnetic fields and placed in an open
field surrounded by high walls. So that the ambulatory activity of
the animals could be measured, the floor of the field was marked
into squares of equal areas. The number of squares traversed within
an experimentally designated period of time was used as the animal's
activity score.

Rats that had been exposed to the 0.5 Hz magnetic fields were
observed to traverse an average of twice as many squares as rats
that had been exposed to control (background geomagnetic field) conditions. Statistical analyses indicate that the field exposure could
explain about 10 percent of the variance in numbers of squares traversed. In comparison, the normal decrease in activity as a function
of number of times tested in the field accounted for about 50 percent
of the variability.

The open field test as well as general activity measures are
crude, diffuse measures. They are helpful as initial behavioral

batteries for diagnosing the types of responses most influenced by an experimental treatment. Alterations in numbers of fecal boluses suggest emotional (autonomic) changes. If the test rat falls over, especially backwards, when rearing, vestibular portions of the brain are implicated. However, subtle quantitative differences in open field behaviors are influenced by a wide range of noncontrolled stimuli from the environment and especially from the very proximal experimenter.

Operant measurements, the instrumental core of behavioral psychology, allow greater precision, control, and predictability of the phenomenon detected. Assuming that the effects of ELF magnetic fields would be weak at most, a measure that normally tests the upper limits of the albino rat's inhibitory capacity was selected: the DRL or differential reinforcement of low rates of responding schedule. Many times in experimental psychology, weak effects have been masked or missed because the animal maintained alternative responses to compensate for any subtle effects from the environmental manipulation. The mammal's capacity to compensate, almost to the point of normality, for cerebellar damage is a case in point.

The DRL schedule requires the test animal to postpone responding for an experimentally determined duration of time. If the schedule is a DRL six-second contingency, the rat must delay responding for at least six seconds in order to, for example, press a lever and receive food reinforcement. If the animal responds in less than six seconds after the previous lever press and reinforcement, the option for reinforcement is postponed another six seconds. Theoretically, if the animal did not inhibit responding and learn the task, it would never receive reinforcement. Stimuli interfering with inhibitory behaviors, such as stimulants or even respiratory ailments, delay acquisition times.

The measure of learning is usually some efficiency measure, such as the number of responses displayed per reinforcements received. Clearly, if the ratio is very high, the rat displayed many more responses than required to receive reinforcements. In one study, rats averaging response/reinforcement ratios of about 3 were placed in different intensity 0.5 Hz fields or control conditions for three days during which they were not trained. When tested after the exposure period, the control rats showed ratios between 2 and 3 (typical of consolidation), while those exposed to 0.8 to 1.0 gauss fluctuations showed similar scores. Rats that had been exposed to from 3 gauss to 30 gauss variations displayed scores from 4.5 to 6.0, but recovered to baseline values within one day.

A similar but persistent effect was noted with adult rats that had been tested for several weeks on a DRL 12-second task, in order to obtain maximum response/reinforcement efficiency ratios (between

1.1 and 1.3), and housed between sessions in sham field conditions. As an example of a typical trend, a rat that showed ratios between 1.2 and 1.3 during baseline (no 0.5 Hz fields between testing), displayed a ratio of 1.6 on the test day following one day of 0.5 Hz, 1 gauss variations. The ratios returned to baseline within two days even though the field was maintained.

When the field was removed for several days, there were no apparent disruptions in the rat's response ratios. Baseline values ranged between 1.1 and 1.2. When the 0.5 Hz field was introduced during the housing period again, the ratios increased on the following test day to 1.4, but returned to baseline within two to three days. A third sequence produced similar but weaker effects. Habituation to magnetic stimuli presented too frequently is a problem rarely recognized.

The continuous controversy and contradiction within ELF magnetic field research is linked to the weak and probably synergistic nature of ELF effects in general. As the many physiological experiments in this laboratory have demonstrated, statistically significant wet weight or structural changes in selected tissue (with an effect that can explain not more than 10 percent of total variability of the measure) are masked by a variety of typical experimental variables, such as housing, background temperature, and humidity, before the magnetic field exposures. Sloppy experimental designs characteristic of hit-and-miss researchers ignore weak but potentially important biobehavioral effects. No amount of covarying out the sources of variability compensates for sloppy methodology.

A major problem of ELF research is maximum reduction of extraneous sources of variability. When this is accomplished, trends are clear, even at the weakest field intensities. For example, rats exposed for five days to sham fields or to 0.5 Hz magnetic field variations of 10 gamma (10^{-8} T or 1 milligauss), 100 gamma, or 1,000 gamma displayed triglycerides blood contents of 186 milligrams/deciliter (mg/dl), 191 mg/dl, 204 mg/dl and 238 mg/dl, respectively, but with great variability. As one would expect from systems theory, the more sensitive systems are also labile. Triglycerides in the blood, for example, also respond to almost anything, from noise levels to the constituents of rat chow.

Duration of exposure appears to be a critical factor in ELF studies. In human subjects, acute, short-term exposure to ELF fields does not appear to induce clearly differentiable effects. On the other hand, long-term exposure, in the order of days, to ELF fields produces changes that are not only internally consistent but also comparable with correlational data. The works of Wever are succinct examples.

When rats are exposed during their entire prenatal development or perinatally (for about five days near birth) to 0.5 Hz fields, persistent and long-term alterations in behavior and physiology are apparent. Adult rats that had been exposed perinatally to 0.5 Hz fields of 1 gauss demonstrated from 5 percent to 15 percent increased testicle weights, 20 percent decreased thymus weights, and 20 percent suppressions in serum glutamate oxaloacetic transaminase activity and lactate dehydrogenase activity (Persinger, Carrey, Lafrenière, & Mazzuchin, 1978). The behaviors of these animals, when tested on a variety of operant tasks, indicated that they were more "emotional." Depending upon the measure, perinatally and prenatally exposed rats have displayed from 20 percent to 50 percent greater "emotional" scores.

There is no argument that long-term exposures to a variety of environmental agents produce detectable changes as a function of time. Technically, of course, continuous exposure of rats during prenatal development to 0.5 Hz magnetic fields may not be immediately generalized to human subjects, whose typical gestation period is 12 times as long. Comparable conditions would require some ELF magnetic condition continually for nine months. Short-term periods, such as geomagnetic/weather disturbances, have peak-trough times too short to be effective. Periods with temporal increments of about one year and greater peak-trough times, such as the solar-geomagnetic couplings of six or eleven years, would be required.

The second type of animal/tissue experiments have involved the use of electric fields at the popular 10 Hz or at commercial frequencies between 50 Hz and 75 Hz. In a long exhaustive series of experiments involving monkeys, rats, and pigeons, John de Lorge and his colleagues (1974) at the Naval Aerospace Medical Research Center have used electric fields between 1 V/m and 30 V/m with magnetic variation components in the order of 1 gauss. They have used almost all the basic measurement procedures from operant analyses, but have found no reliable or consistent effects from ELF field exposures.

At present, there are no clear reasons why such discrepancies should exist between studies with rats exposed to 0.5 Hz magnetic fields and the results collected by the de Lorge group. Several procedural differences exist, which, if demonstrated simultaneously, may explain the apparent contradictions. For example, de Lorge placed his test animals into the ELF field situation and removed them during nontest periods, the exact opposite of the 0.5 Hz studies. Secondly, most of the exposure periods in the de Lorge studies were about one hour or so per day. Perhaps, if the Wever and Ranscht-Froemsdorff data have application here, much longer periods of exposure are required.

The research of Lang and Altman at the University of Saar-brueken (Altman, 1974), on the other hand, lends strong support to the work of Ranscht-Froemsdorff, Ludwig, and Wever. Their basic procedure has involved three exposure conditions: a normal environment, a Faraday cage environment in which ambient atmospheric ELF signals were shielded, and a Faraday cage environment within which artificial 10 Hz square waves were produced.

After about two weeks of exposure to these conditions, the hematocrit values (a measure of the relative volume of blood cells in the whole volume of blood) fell from 43.9 in normal mice to 32.2 in the Faraday-housed mice, who also showed a 30 percent increase in water retention. Similarly, sodium concentration increased from 60 milliequivalents (meqv.) to 68 meqv. in whole blood and from 6.2 meqv. to 9.6 meqv. in red blood cells, while potassium levels dropped in a proportional manner. Altman suggests that the effects of Faraday conditions increased the colloid osmotic pressure within the animals' volume, resulting in increased water retention and an associated drop in hematocrit.

A return toward control values occurred in mice housed in the Faraday cages within which 10 Hz fields were applied. Hematocrit values rose to 36, still lower than normal, while sodium concentration in the blood dropped to 58.9 meqv. and to 4.4 meqv. in the red blood cells, essentially normal conditions.

At face value, the results, based upon at least 2,000 animal samples, confirm the correlational data involving hematological difficulties and ambient ELF fields. A drop in hematocrit values and, hence, blood cells per volume of blood would certainly hinder or contribute to the blood clotting time variability. Intensification of water retention by possible barometric pressure changes could further complicate the problem.

Lang and Altman's data are persuasive, until one determines the intensities of the square wave 10 Hz field. In most of their studies, values exceeded 1,000 V/m. So, when are experimental intensities no longer comparable with natural values? When is an experimentally tractable paradigm with differentiable results unrealistic to correlational comparisons with vague patterns? This is a scientist's Gordian knot; the solution will require more than a swift, single slice of methodology.

Experiments involving simulated atmospherics ("sferics") patterns and intensities, the ultimate biometeorological design, have been limited in nonhuman studies to tissue measurements. Impressed by the significant correlation between atmospherics values and day to day variations in tissue respiration, Lotmar, Ranscht-Froemsdorff, and Weise (1969) constructed a shielded room with steel reinforced concrete walls lined with 2 mm steel sheets. A

Warburg apparatus, used to measure tissue respiration, was placed in the middle of the room between 2 m^2 plates separated by 2 mm as well as within coils.

Fresh mouse liver tissue was used in the reaction vessels. After obtaining control values at 15 minutes and 45 minutes after initiation, the tissue was exposed for 30 minutes to artificially generated bad weather impulses (from 30 to 100 Hz pulse frequencies greater than 100 mV/m with carrier frequencies between 10 kHz and 100 kHz) or to good weather (from 1 Hz to 3 Hz less than 10 mV/m via 10 kHz sources) impulses. Whereas the bad weather fields decreased respiration rate of the liver tissue by 42 percent $\pm$ 3 percent relative to control values, the good weather effects were not effective.

MAGNETISM MANIA IN THEORY

A variety of theories and explanations have been published regarding the mechanisms by which magnetic fields influence the behavioral system. Most of these theories have been derived from physical chemistry levels of discourse. The basic assumption appears to be, "We can describe the magnetic field to be a simple phenomenon of physics, hence the biomechanism must be described in a similar manner."

This section will review, very briefly, the general types of theories used to describe or explain the putative biobehavioral effects. Most of these theories have been limited by their general nature and the failure to clearly generate testable hypotheses. More detailed descriptions and varieties of theories are reported elsewhere (Persinger, 1974; Persinger, Ludwig, and Ossenkopp, 1973).

One popular model from physical chemistry involves the oxygen atom and its paramagnetic qualities. Paramagnetic materials are attracted to, while diamagnetic substances are repelled from, a magnetic field. Oxygen changes from a paramagnetic to diamagnetic condition when it enters into a reaction or becomes a part of an organic molecule.

Since intense static magnetic fields in the order of 1,000 gauss are well known to retain the paramagnetic qualities of oxygen, such fields will interfere with oxidative reactions. Presumably, the application of static magnetic fields will reduce the amount of oxygen entering the reaction relative to the amount that would have been available without the magnetic field.

Some theorists have applied this concept to both very weak intensity static magnetic fields and to ELF magnetic fields. Using the "maintained paramagnetic possibility," they assume that a consequent effect of magnetic field presence is relative hypoxia. This hypoxia would be statistical and due to magnetic field retardation of oxygen changing from a paramagnetic to a diamagnetic (for example, oxyhemoglobin) condition.

Given this assumption, a number of qualitative hypotheses can be generated. First, since oxygen is carried to tissue through the blood while bound to hemoglobin, this type of mechanism should be associated with altered concentrations and/or activities of hemoglobin. Reactions preceding and following the hemoglobin production should be affected as well.

The second popular theory involves some alterations within that enigmatic interface between life and death: the cell membrane. Since the cell membrane exists as a special liquid crystal, slight alterations in the structure would induce statistical increases in the number of ion entrance points. Small amounts of sodium could leak in, and small amounts of potassium could seep out.

The general mechanism for this theory involves alteration of very special conditions: selective permeability in the cell membrane. This alteration is assumed to be completed by a variety of mechanisms, including simple electron flow, cascading release of charges (such as a transistor), stimulation of miniature postsynaptic potentials, or various superconducting options, such as the Josephson junction. Current physical theory appears to determine the popular choice.

In all situations, the magnetic field is assumed to be a type of trigger that induces a structural change in the liquid crystal matrix of the cell membrane. In the neuron, the slight change in ion entrance points could alter the probability of firing within large clusters of neurons. In the red blood cell, the slight alteration could influence the relative concentrations of sodium and potassium as well as the fragility of the membrane.

The model is heavily dependent upon some type of cell membrane heterogeneity concept. If all cells in the body responded similarly to the applied magnetic fields, the changes would produce similar alterations in all systems. Although the absolute baseline may change, the relative relationships between cell systems would remain the same.

Usually specific loci within the body space that show special fragility of membrane property or electrical congruence are selected. Parts of the brain that normally generate 10 Hz fields, such as mid-line nuclei of the thalamus or the occipital cortex (alpha rhythm), are popular choices. Parts of the brain that involve extreme electrical instability, such as the temporal lobe, are selected as well.

The unstable options appear attractive since the temporal lobe and the hippocampal-amygdaloid complex, perhaps the most unstable part of the human brain, are involved in weather-related symptoms. Statistical stimulation of these areas would not necessarily produce obvious seizures. Instead, increased probabilities of paranoid thoughts or altered intensities of concentration could result.

Applied to the erythrocyte, enhanced fragilities of the membrane would produce greater numbers of cellular fragmentations. Consequently, blood-related organs would demonstrate increased activity. Spleen and other phagocytic organs would increase in size and activity as the increased proportion of fragile cells were destroyed within their boundaries.

The third basic model has been a "substance that is everywhere" concept. This concept must be taken with the greatest precaution since it can degenerate into magical thought very quickly. One could be left with a phlogiston argument, reminiscent of times past.

Likely candidates are the very versatile components of the connective tissue system: the protein polysaccharides. These structures, such as the chondroitin sulfates, hyaluronic acid, and heparins, are found throughout the body within the cells and within extracellular fluid. They contain complex meshes of charges that could quickly alter aqueous reactions since they loosely sequester water molecules.

Stimulation of these systems would be expected to alter fluid balance, in general, at the chemical level. Organs that contain a disproportionate amount of these substances, such as the atria of the heart, the joints, and the humors of the eye, might show enhanced water accumulation. Excessive edema could precipitate heart failure, alterations in blood pressure, arthritic complaints, and ocular pain.

Progressive alteration of these substances in critical interfaces, such as between the blood and brain or some other tissue, could lead to deleterious consequences. Erosion of the elastin fibers within the arteries and arterioles could increase their fragility by changing the elastic nature of the walls. Sudden constriction or dilation could produce damage.

Within the brain, these substances could be important as well. W. R. Adey suggests that the effective membrane boundary exceeds beyond the normal 100 angstrom units (Å) of the cell. Instead, the functional width of the neuron extends as far as 1,000 Å into extracellular space. This added dimension is composed of a fragile fabric of protein (muco) polysaccharides, to which calcium, sodium, and other ions are bound.

Presumably, the small currents generated by the passage of the magnetic flux lines through the body volume would be sufficient to alter the electrical balance and characteristics of the protein polysaccharides. Interestingly, both theoretical and empirical measurements indicate that these chemicals would be sensitive to extremely low current densities of less than a nanoamp (10^{-9} A).

Assuming very conservative estimates of resistivity (10 ohm-meters), a 1 Hz field of 10^{-4} T (1 gauss) variation intensity would induce currents of a nanoamp. Certainly, localized currents of greater magnitude could occur in clusters of chemicals that display very marked increased electron availability or conductivity, such as the protein polysaccharides existing in the liquid crystal phase (Persinger, Lafrenière, and Ossenkopp, 1974).

Confirmation of any of the magnetic field hypotheses will require much work. Most biomagnetic theories are postulated almost as an afterthought to fill the requirements of the discussion section within technical journals. Since so many of the speculations can explain the data equally as well, large matrixes of data will be required before the most likely model can be discriminated.

10

SOLAR, LUNAR, AND POTENTIAL PLANETARY CORRELATIONS WITH HUMANS AND THE WEATHER MATRIX

All energy involved with weather processes ultimately originates from the sun. This star generates a wide range of electromagnetic energies, from increments directly experienced as heat or light to forms with indirect influence, such as X rays and radio waves. One would expect, in theory, that variations, either periodic or random, in the sun's energetic outputs should influence earth weather. At the practical perspective, the unknown temporal delays between solar changes and weather alterations and the multitude of energy exchange mechanisms that exist between the sun and the earth's surface present a very difficult quantitative problem.

Two other extraterrestrial factors may account for a significant, although weak, influence upon humans and the weather matrix: the moon and some of the planets. Lunar effects are a theoretical necessity since the moon is sufficiently close to the earth to induce tidal alterations. In addition, the moon periodically passes between the sun and earth (the new moon) and distal to the earth on the side away from the sun (the full moon). Such movements not only influence the movement of the solar wind, but also alter the dynamics of any energetic particles trapped in the upper magnetosphere.

The second extraterrestrial factor is also a theoretical requirement, but should be maintained as a very probationary condition. Some recent observations indicate that when the planets Mercury or Venus pass between the earth and sun (inferior conjunctions), statistically significant alterations in geomagnetic disturbances are noted, presumably through some as yet unspecified alteration in the expansion of the solar wind. A second group of observations and calculations suggest that Mercury, Venus, Earth, and Jupiter (because of its mass) could, during critical solar periods, produce sufficient tidal forces to alter solar activity.

All of these processes—the sun, the moon, and potential planetary contributions—are far away from the relationship between
humans and weather. As should be evident by now, the objective
determination of which variable or variables within the weather
matrix influence the variety of different human behaviors is already
difficult. The addition of processes, removed by more than three
levels of discourse and demonstrated by correlations with correlations,
is a tenuous procedure. An objective scientist must be cautious but
not exclusive.

THE SUN

The sun would be classified in astronomical categories as an
unimpressive G2 star. This star is approximately 864,000 miles in
diameter (or about 100 times wider than the earth) and displays a
mass of about 10^{33} gms. Although the sun rotates, its different parts
do not display the same periods. Whereas rotation time at the equator
averages about 27 days, periods exceeding 30 days are evident at the
solar poles.

Like other stars, the sun's energy is derived from the trans-
mutation or fusion of hydrogen into helium. The source of the energy
originates from within the sun's central core, which constitutes less
than 1 percent of the total volume but about 40 percent of the total
mass. Here the temperature is estimated to exceed 15 million
degrees.

The energy from the sun's central core moves outward in the
form of X rays and gamma rays to distances of about 0.7 the radius,
where radiation of energy is no longer possible due to physical
limitations. At this distance, the temperature has dropped to an
estimated 200,000 degrees. Analogous to water boiling in a kettle,
solar material carries the energy to the surface. This region is
called the convective zone.

The visible surface of the sun, the photosphere, constitutes
the outer interface to the convective zone. Typical photospheric
temperatures average about 6,000°K, while matter densities average
values about one-thousandth that of earth air at sea level. Photons
emitted beneath the photosphere are stopped by internal processes,
while conditions above the surface are too rarified to allow efficient
release of these particles. Since the photosphere is only from about
200 miles to 300 miles thick and without clear boundaries, the majority
of solar photons that reach earth are emitted from a shell with a
thickness less than one-thousandth of the solar diameter.

The ultrastructure of the solar surface does not display the
apparent smoothness seen with low magnitude photography. In white

light, the normal solar surface displays a granulated appearance. Various metaphors have been used to describe the solar complexion, such as rice grains, granules, or cobblestones, depending upon the historical background of the astronomer. The individual "cells" or granulations have varied polygonal shapes with diameters primarily between 300 miles and 1,000 miles and with mean lifetimes of a few minutes. The granules have been interpreted as consequences of convective forces from heated gases rising to the surface from hotter invisible layers below the photosphere.

Above the photosphere, the solar atmosphere has two layers—the chromosphere and the corona—both of which are invisible in ordinary (white) light. The chromosphere extends upward and displays a distinct division at about 2,500 miles above the surface. Below this height and above the solar surface, the hydrogen is nearly all neutral and "cool." Above this zone to distances reaching the corona, some 13,000 miles above the photosphere, the hydrogen gas is very hot and almost completely ionized.

When the sun is photographed through special filters that tune out the whole electromagnetic spectrum except for the H α line, the portion of the spectrum associated with the emission of hydrogen, otherwise invisible structures are seen in the chromosphere. Bright short-lived jets or spicules are seen in H α photographs of the solar edge or limb. At least 100,000 spicules have been estimated to be present at any moment on the solar surface, with lifetimes of about two minutes and maximum visible heights of about 5,000 miles above the photosphere. They are assumed to be related to the short-lived granulations on the solar surface and are probably an important factor in the transportation of energy upward toward the corona.

The corona is the outer solar atmosphere, extending outward from the chromosphere, millions of miles above the center. Kinetic temperatures, that is the average kinetic energy of electron and atomic motion, are extraordinarily high, in the order of 1 million °K. Despite these high temperatures, the low density of matter at these distances (only about 10^5 atoms/cc) keep the radiation energy much lower. Presumably, the energy required to maintain the high coronal temperatures is derived from the continual intrusion of chromospheric spicules that are, in turn, related to the granules of the photosphere.

Ionized hydrogen, which is the primary source of protons (hydrogen nuclei shorn of electrons), electrons (that have escaped from the hydrogen nuclei), and small amounts of other atomic fragments, stream away from the outer coronal regions when the number density and free space reach optimal values. The expanding corona constitutes the "solar wind," which, during quiet solar periods, averages about 10 outbound coronal corpuscles per cc. As the earth orbits around the sun, it moves through a medium of ultrararefied hot ionized hydrogen. Solar-earth transit time of the quiet condition "wind" is from about 4 days to 5 days.

Monochromatic Sunspot-related Phenomena

Photographs that filter out all incoming sunlight except for tiny wavelength intervals at the red absorption line of hydrogen (Hαline) or the line associated with ionized calcium, reveal a number of sunspot-related energetic phenomena. Faculae are bright granulated patches surrounding sunspots near the edge of the solar disk, with lifetimes of about two weeks. Dark, narrow, snakelike streaks or filaments are often seen near sunspots and faculae. When seen at the edge of the sun, filaments appear bright and red, at which time they are called prominences.

Prominences are relatively long-lived phenomenon, lasting for several solar rotations. Only about 3,000 miles thick, these curtainlike parcels of gas stretch well up to from 20,000 miles to 50,000 miles into the solar corona. Over time, they are distended by differential rotation speeds between the solar surface and the upper boundaries. Consequently, they curve above the solar surface in an east-west direction. Prominences are popular paired-associations of solar phenomena. Film clips of these photographically impressive phenomena are seen in documentaries and science fiction movies.

The Sunspot Cycle

The occurrence of the 11-year sunspot cycle is probably one of the most well known time-varying characteristics of the sun. Sunspots have been easy to record, since they are quite visible in white light, even with relatively weak telescopes. The Wolf numbers and Zurich numbers, R, are a function of some combination of the total number of spots and the number of spot groups. A closely related measure is the sunspot area over the entire solar surface. The sunspot cycle is closely related to solar activity cycles, defined in terms of flare-related characteristics. This close temporal proximity, as Smith and Gottlieb (1974) have indicated, has resulted in the erroneous attribution of some solar phenomenon to the sunspot cycles.

The diameter of a spot may range from 1,000 miles to 30,000 miles. Groups of spots can attain lengths of more than 125,000 miles and contain up to 100 spots. Whereas a single spot lasts for about 1 day, the maximum size of a group of spots generally requires about 10 days. Spot groups rarely survive more than one solar rotation (about 28 days). Exceptionally large groups of spots may survive as many as five solar rotations.

In a technical sense, sunspots are not really black. This apparent darkness is due to the intense contrast between the cooler, central area (umbra) of the sunspot, which is about 4,500° K, and the much hotter and brighter photosphere, which is about 6,000° K. Marked reductions in temperature within sunspots have been attributed to intense magnetic fields generated within/around them. Field intensities approaching 10,000 gauss have been estimated by using well-accepted inference procedures. These massive and tangled magnetic flux lines appear to behave like "magnetic bottles" that contain and delay the upward moving gases long enough to allow temperature reduction or to prevent temperature increases.

The spot itself is a short-termed, secondary manifestation of a more profound solar magnetic process. Magnetic fields with dimensions of a sunspot have been known to last well over a century on the solar surface. The strength of a local magnetic field rises quickly after the spot has been formed, remains constant for a time, and then rapidly decreases in intensity before the spot disappears.

Sunspots have been interpreted generally as local intensifications of the general magnetic field of the sun. Some theorists suggest these intensifications are normal displays of solar magnetic variability; others have argued that differential rotations between the polar and equatorial regions are particularly responsible. In the latter situation, the differential rotations would induce vortices within the sun (whirlpool-like changes) where the general solar flux lines are tangled and twisted in a concentrated space. The two ideas are not necessarily mutually exclusive.

Flares: Short-term Solar Flux Variations

Flares are highly energetic phenomena with lifetimes in the order of one minute. Although they are traditionally defined by the enhancement of Hα radiation, flux increases for X rays and the far ultraviolet range are quite apparent as well. Optical flares are classified according to the area and brightness of Hα radiation. The importance rating of a flare is a function of the solar area involved and ranges, from decreasing to increasing importance, from 1 to 4.

Flares are most numerous during sunspot maximum. During the maximum, flares of importance 1 or greater appear about every two hours. Of all flares recorded during the maximums of 1955/1961, 89 percent were of importance 1, 10 percent were of importance 2, and less than 1 percent were 3 or greater. Apparently, although the number of flares changes with sunspot cycle, the proportion of different importance flares does not.

The power output in X-ray, extreme ultraviolet, and Hα spectra may increase by several orders of magnitude during a flare. Soft X-ray bursts, for example, have a rise time (the duration from baseline levels to peak levels of output) close to 4 minutes and decay times of 12 minutes. Similar short-termed enhancements occur in the meter-wavelength continuum. For example, during a flare on September 27, 1969, a train of 50 sharp pulses with intervals between pulses of 2.5 seconds (ELF pulses) was received in the 100 to 200 MHz band for about 2.5 minutes. These periodic pulses began about 25 minutes after the start of a large flare.

The Solar Constant

The total irradiation output from the sun was traditionally assumed to be constant. Early calculations indicated that the output was in the order of from 1,350 to 1,400 W (watts)/m^2 or, when calculated to the distance of the earth, from 1.9 to 2.0 calories/cm^2/minute. Later, some theorists argued that the constant varied in the order of 1 percent as a function of solar activity. On a simplistic level, one would expect that even small variations in the energetic output from the sun might influence meteorological and biological phenomena. One can begin to answer this problem by determining whether the electromagnetic spectrum generated by the sun varies periodically and if the alteration exists, what portions of the solar spectrum, for example, ultraviolet or heat (infrared), vary the most.

By knowing the surface temperature of a star, one can calculate roughly the wavelength at which the most energy will be emitted. According to Wien's law, the wavelength with the greatest incidence and energy is equal to 0.29 cm-degrees divided by the temperature in degrees K. For the sun ($\sim 5,500°$K), the maximum output energy is emitted around 5.3×10^{-5} cm or 5,300 A or what we see as yellow yellowish-white. One would not expect significant variations of flux output at peak wavelengths, since this would imply unusual core instability. Output intensity variations at peripheral wavelengths comprising not more than 1 percent of the total power output would be expected from several physical and statistical models.

Quantitative analyses support the general calculation. In a review paper that integrated data from many sources, Smith and Gottlieb (1974) showed the relative contributions of various wavelengths between 2 A (10^{-8} m) and 20,000 microns. Wavelengths below 5,000 A contributed only 22 percent to the total output from the sun. Wavelengths above 16,000 A contributed about 10 percent. The majority of the power output lies in wavelengths between 5,000 A and 16,000 A.

After exhaustive analyses, Smith and Gottlieb concluded that the major determinants of the solar constant, the near ultraviolet, visible, and infrared wavelengths, do not vary significantly with solar activity. The most significant cause for variation in solar flux is the varying distance between the earth and sun. This one-year period swamps any of the small variations in these components due to flare, solar day, or 11-year cycles. However, variations in X-ray and extreme ultraviolet portions of the solar flux, although very small in terms of total energy output, may vary several orders of magnitudes during major flares.

Principles of Solar-Terrestrial Interaction

Excluding the possibility of complex functions or nonlinear relationships, one would expect short-term solar fluctuations, with durations in the order of from minutes to days, to be most involved with weather changes. The gigantic sinusoidal changes in energy that reach the earth as a function of solar-earth distance during the year are important for seasonal alterations. But these factors are too long and too gross to overlap with the lifetime of weather processes. (One could imagine the turbulent nature of the earth's surface if solar changes in the order of a season occurred within a week.)

Two general processes appear to be likely interfaces between short-termed solar activities, which may increase in number as a function of sunspot cycle, and terrestrial weather: the solar sector contacts and flares. As mentioned in Chapter 9, energetic release of particles by flares into the solar wind can dump massive amounts of particles into the upper magnetosphere. These quick massive injections of particles alter ionospheric processes, "causing" breakdowns in radio communications as well as global magnetic storms. Importance 3 B flares can emit energies in the order of 2×10^{30} ergs, although only a small portion of this reaches the earth.

The determination of which constituents associated with flares contribute to weather changes or whether different constituents, such as X rays, proton bursts, or radiowaves, contribute in different ways to alter weather have not been completed. Although corpuscular contributions, such as ionized particles, would be theoretically more important than ultraviolet frequencies in the upper ionosphere, the influence of the latter flare correlates upon lower atmospheric constituents, such as the ozone layer in the stratosphere, is less well-understood. Because of the close proximity of different flare electromagnetic-particulate phenomena in time, isolation of "what is contributing to what" will be difficult.

Whereas flares ride as ripples on the solar activity 11-year cycle, a more normal periodicity occurs between the interface of the earth's magnetic field and the solar interplanetary magnetic field. As seen in Figure 20, the solar magnetic field is divided into sectors. Each adjacent sector section displays an opposite polarity, which dictates, to some extent, the direction of particle movement. During quiet solar periods, one would expect that terrestrial crossing of the solar sectors, where turbulence should occur, to be associated with weather changes.

Interplanetary magnetic field lines from a sector directed away from the sun can connect most readily with geomagnetic flux lines directed into the north polar cap, whereas interplanetary magnetic lines directed toward the sun should most readily connect with geomagnetic flux lines emerging from the southern cap. The periodicity and soluble geometry of sector sections make these phenomena excellent test areas to demonstrate solar-weather relationships.

Solar-Weather Correlations

The relationship between solar activity and weather is simple in principle but almost impossible in detail. A multitude of solar-weather correlations have been reported over the last 20 years. These studies have compared weather and climate with a wide variety of solar periodicities, from short-termed flare relationships to the more cumbersome 27- or 28-days and ~11-year cycles. By far the most superior review of these correlations and the pitfalls of analyses have been reported in a technical article by Pittock (1978).

Correlations between weather, as a phenomenon, and solar activity are surprisingly few. Most of the weather-related cross-references in the literature usually involve a singular measure of weather, such as rainfall, or refer exclusively to processes in the upper ionosphere. No doubt, as one would expect, there are very large and reliable correlations between solar activity and myriad processes in the upper ionosphere, the primary interface between the atmosphere and the solar wind. However, there is a very long conceptual and quantitative gap between ionospheric processes and local weather changes.

Relationships between weather and the earth's passage through a sector boundary of the solar (interplanetary) magnetic field have been reviewed by J. M. Wilcox (1976, 1975). When the interplanetary field was away from the sun, a process that would facilitate "connection" through the northern polar regions, average barometric pressure readings were 1,011.0 mb in the northern stations and 986.2 mb in the southern polar stations. When the interplanetary field was toward

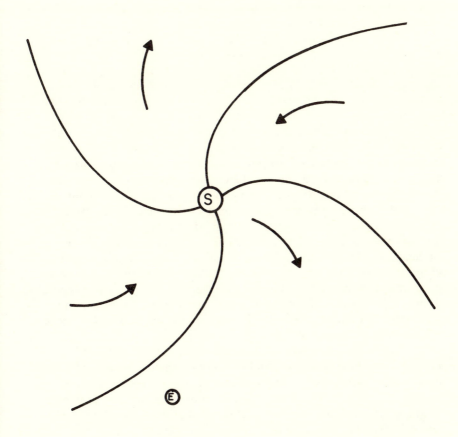

Figure 20   Sector sections of the solar magnetic field.

Note: S and E stand for Sun and Earth respectively.

the sun, a condition that would facilitate southern pole connection,
the southern stations showed a relative lower pressure measure
(982. 7 mb), whereas the northern stations showed relatively higher
pressures (1,016.3 mb).

According to Wilcox, low pressure troughs in the northern
hemisphere winter display decreased measures of strength during a
few days near the times when sector boundaries are carried past the
earth by the solar wind. The amplitude of the effect is about 10 percent of the mean. Similarly, thunderstorms, a phenomenon important
for the maintenance of equilibrium between ionospheric and ground

charge, appear to increase when the earth is crossing from an "away" (from the sun) sector into a "toward" sector.

As mentioned in Chapter 9, the concomitant alterations in geomagnetic activity per se add serious confounding factors to clear discrimination of mechanism. In the days before a boundary, the geomagnetic activity (K_p index) displays a monotonic decline to a minimum about one day before the boundary. Geomagnetic activity then rises to a peak, about twice trough values, from one to two days after the boundary passage and then declines. Several weather phenomena, from significant increases in westerly winds to alterations in atmospheric pressures are correlated with geomagnetic disturbances (in general) three to four days before. The effects of geomagnetic disturbances and tropospheric changes are greater in the winter.

Not surprisingly, the consequences of intense flares evoke not only massive geomagnetic disturbances, but also alterations in the same weather variables previously noted. Significant alterations in the height of given millibar levels above the earth's surface occur within 24 hours of a flare, although significant changes do not occur at ground level until about three days after the flare. Global thunderstorm activity can increase by from 50 percent to 70 percent on the third and fourth days after a Hα flare. Again, the effects are most evident during winter seasons.

Long-term correlations between solar activity and weather become less valid since the numbers of confounding factors enter the problem in a nonlinear manner. There is good evidence to support the contribution of 27- through 29-day solar rotation time to geomagnetic disturbance, a factor not unsuspected considering the solar sector correlations. As pointed out by the work of Fraser-Smith (1973), the isolation of solar contributions is more difficult than simply aligning data lists.

Certainly, the most popular correlation efforts have been between sunspot cycles and weather or, more accurately, climate. Climatic or long-term alterations in average environmental values, are well known, but appear to be very situation-specific. The growing season in parts of Scotland is very closely coupled with sunspot cycle, although it is not clear how many areas would be correlated by chance alone due to intrinsic periods.

Thunderstorm activity has been shown to be significantly correlated with the 11-year period across four solar cycles. As the annual mean sunspot numbers increase from lows of about 25 to highs of 100 or more during sunspot maximums, annual lightning incidence measures range from lows of about 60 to peaks of about 120. The thunderstorm activity is in phase with the solar cycle with an amplitude of ±30 percent of the mean.

To be skeptical does not mean to be exclusive. This point is best supported in a paper by Blanco and Catalano (1975). These authors correlated air temperatures in Catania, Italy for the period from 1817 to 1970 with the sunspot number R. The temperature and the sunspot number, both averaged over the 11-year solar cycle, gave a correlation of $r = 0.5$. Higher mean temperatures were associated with more active solar cycles, while lower mean temperatures were associated with cycles of lower mean solar activity. Although interesting, their major finding was that no reliable correlations existed for time scales of less than a solar cycle.

Solar Activity and Living Systems on Earth

Without the totality of visible, near ultraviolet, and infrared wavelengths from the sun, there would be no life on earth, as defined by humans. There would be no weather. The variations in solar flux that reach earth as a function of solar distance, although relatively small in terms of absolute magnitude, are sufficient to induce (in combination with changes in the earth axis angle) the massive changes in season. Seasonal effects are clear baseline factors upon which much smaller weather transients are superimposed.

Although direct effects of solar radiation are clear, secondary and tertiary influences upon living systems have not been properly evaluated. The subtle psychological consequences of sunny days compared with cloudy days are accepted in principle but have not been demonstrated in particular. In all instances, these proximal factors should be considered to be orders of magnitude greater in day-to-day importance than the direct consequences of sunspot activity or flares.

Correlations between flares or sunspot numbers and biobehavioral measurements have occurred repeatedly. In fact, in some arenas of science, they have become the initiates' joke into the finer understandings of statistics. The correlation between sunspot numbers and wars has been used to indicate how anything can be correlated with anything. The lesson, like many tricky areas of borderline science, is neither all correct nor totally erroneous. A persistent correlation that is replicated many times is indeed a reliable statistic; it does not mean, by its reliability, that the two variables are causally related.

The Ubiquitous 11-year Cycle

The pervasive nature of the 11-year sunspot cycle in terrestrial events is a perplexing problem. Two examples will be given here to demonstrate the depth of the relationship. In the first, Barber (1962) noted a marked alteration in the effective capacity of the storage battery system over the years. Using detailed measurements collected since 1920, he correlated the mean annual potentials with sunspot numbers over four 11-year cycles. His results indicated a correlation of $r = 0.56$ between the voltage measurement and sunspot numbers when the former's measurements were lagged 1.5 years behind sunspot numbers. Apparently, the effective capacity of the system diminished by 30 percent during the first to second year after the sunspot maximums.

In the second instance, the controversial variance in the visibility of Jupiter's red spot has been linked to solar activity. Basu (1969) reported a correlation coefficient of $r = 0.57$ for 57 pairs of observations representing the years 1894 to 1945. However, after 1947, the relationship became less apparent. Basu interprets this relationship as indicative of some local variation in the density of the interstellar medium. However, the conceptual distance between terrestrial events and alterations in interstellar medium is reflective of the magnitude of these distances.

Mechanisms of Solar-Weather Interaction

One would suspect from the general data patterns presented that a major, although not singular, mechanism of solar-earth interaction would be particle injection. D. M. Willis (1976) has published an eloquent argument against the direct transfer of magnetospheric energy acquired through solar wind/flare interaction to the lower atmosphere. The general metaphor for the latter concept is that agitation of outer geomagnetic flux lines could transfer energy into the lower atmosphere.

A group of German scientists led by Reinhold Reiter (1976) has demonstrated an important empirical link between solar events and weather processes. Their data show increased frequencies of stratospheric injections of charged particles into the troposphere following solar events. These injections are associated with a maximum atmospheric potential gradient within a few days after Hα flares.

To demonstrate the relationship, Reiter's group, like Wilcox, has used statistical techniques that allow the separation of signal from noise. Since routine correlations between atmospheric measures and flares (which are 0, 1 events) are statistically precarious unless one clearly knows the population parameters, they used epoch analyses. In this procedure, phenomena with intense characteristics, such as Hα flares, are used as key days. All days in which Hα flares occurred are given the value of zero. Pre- and postdays for each event are aligned appropriately.

Using radio nuclides, such as berlium, as measurements of stratospheric injections, Reiter found that concentrations of these materials increased to 60 percent over background minimums between the second and fourth day after Hα flares. He concludes that a solar event, such as an Hα flare, increases the probability of stratospheric air injection into the trophosphere by at least 50 percent. Since concentrations of ionospheric materials and stratospheric ozone are well known to be associated with all major periodicities of solar activity, from flares to 11-year cycles, the particle injection hypothesis may be close to verification. The effect of these injections upon ongoing weather activity and biological systems, if any, will no doubt reflect the baseline values of processes occurring at the time of the injections.

THE MOON

The moon is an object approximately 2,160 miles in diameter with a mass ratio 1/80 that of earth. Mean lunar-earth distance is 225,000 miles with variations of 5.6 percent of the mean. During perigee, when the moon is closest to the earth, the lunar distance is 5.6 percent less than the mean distance, and, during apogee, when the moon is furthest from the earth, the distance is 5.6 percent greater than the mean.

Lunar Periodicities and Phases

Several lunar periodicities exist, primarily in the order of 28 days. The orbital or sidereal period of the moon is 27.3 days and is defined as the time taken for the moon to make one circuit through the star canopy. On the other hand, the synodic period of the moon is 29.5 days and is defined by two successive conjunctions (overlappings) with the sun. The discrepancy is primarily due to the added movement of the earth around the sun during the lunar orbital period.

Other periodicities exist. The major cycle shorter than the 28 days series is the daily variation of about 24 hours and 50 minutes. Again the delay over 24 hours is associated with the differential rotation time of the earth with respect to the orbital time of the moon.

Periodicities longer than a lunar month are associated with secondary perturbations in the lunar orbit. The entire axis of the lunar orbit rotates, moving in a counter-clockwise direction. As a result, the point in space where perigee (or apogee) occurs will be one complete revolution every 8.85 years. The two points at which the lunar path crosses (ascends or descends) the ecliptic (the annual pathway of the sun that splits the celestial sphere in half), are called nodes. These lunar nodes are not fixed points but move west along the ecliptic every 19 years. In addition to this orbital rotation, the orbit always "wobbles" with respect to stellar distance. A much weaker amplitude, the effect has a period of 31 years (Dewey, 1959).

The phases of the moon are probably the most well known periodicities observed from the earth's surface. New moon occurs when the moon is in conjunction with the sun. During this time, solar light shines on the lunar hemisphere that we cannot see. Since the new moon is near the sun, in the terrestrial sky, the moon is up all day and down all night. About three days after new moon, the moon has moved east of the sun and therefore rises and sets about three hours after the sun. During this crescent phase, about one-fourth of the illuminated hemisphere can be seen from earth.

Seven days after the new moon, the moon is 90° east of the sun. Consequently, the moon rises about noon, crosses the meridian at about sunset, and sets near midnight. Since one-fourth of the synodic month has elapsed, the moon at this time is called the first quarter. Ten or eleven days after the new moon, three-quarters of the daylight hemisphere can be seen from the earth. Frequently, this phase is called the gibbous moon, which rises in the east about midafternoon.

During full moon, the moon is 180° east of the sun or in opposition. Rays from the sun pass the earth and shine on the hemisphere that is exposed to the earth. Being opposite the sun, the moon rises at sunset above the horizon all night and sets at dawn. Between new and full moon, the moon is said to be waxing.

A reverse series of phases occurs during the waning phase from full moon to new moon. During the last quarter or third quarter, the moon is 270° east—or 90° west—of the sun. The moon rises about midnight and sets at noon. The waning crescent moon rises before dawn and can be seen in the eastern sky a brief period before sunrise.

Biometeorological Relevance

Even if lunar phases were not related to weather, the moon is an important variable in biometeorological phenomena. Folklores of many countries link weather-related pains and ailments with lunar phase, especially full moon. In fact, many of the psychiatric disturbances and general behavioral oddities paired with weather are also paired with lunar phase. However, recent evidence indicates that lunar phase may contribute more to weather but less directly to human behavior, outside of psychological expectancies, than previously suspected.

Lunar Correlations with Atmospheric Phenomena

Reliable correlations between lunar phase and geometeorolog-
ical phenomena have been available for about only 15 years. A not
surprising relationship with lunar phase and geomagnetic disturbances
was reported by Bigg (1963), who used geomagnetic storm data over
an 81-year period. The first-order effect was clearly seen between
full moon and new moon. Whereas the total numbers of storms around
the new moon varied between 180 and 190 over that period; the total
number of storms around the full moon ranged between 220 and 240.
A second-order peak in storms was found a few days before first
quarter. The importance of differentiating qualitatively different
patterns, even though they may still be magnetic phenomena, was
also demonstrated by Bigg (1963). Discriminant analysis indicated
that the new moon decrease was valid for only gradual commencement
storms, whereas sudden-commencement storms showed an increased
frequency at new moon periods.

Hydromagnetic (Hm) emissions, the most outstanding geomagnetic pulsation phenomena observed in the range between 0.2 and 5
Hz (~1 γ amplitudes) do not appear related to geomagnetic variations,
but are correlated with lunar phase. Hm emissions, according to
Fraser-Smith (1969), appear to be generated in the outer equatorial
regions of the magnetosphere and to involve nonthermal protons.
Subsequent downward movement of the signals along a geomagnetic
flux line and then partial reflection from the ionosphere back along
the line result in the repetitive Hm process. This phenomenon shows
a marked enhancement on the day before full moon.

Lunar phase relationships are evident in the upper atmosphere
as well. Several studies have linked variance in meteor rates and
in atmospheric ozone to lunar phase. Both Brierley and Davies (1963)
and Bowen (1963) noted increased meteor rates between the full moon
and last quarter. The latter author reported an increased rate from
155 before the full moon to 175 after the last quarter. A more complicated relationship is apparent for ozone. Whereas ozone concentrations increased during the first and last quarter during the spring,
they decreased with the same lunar phases during the fall. Midsummer and midwinter effects were not apparent (Shah, 1972).

Perhaps the most important atmospheric correlation with lunar
phase, from a biometeorological point of view, is the relationship
between lunar phase and thunderstorm activity. Lethbridge (1970)
analyzed thunderstorm frequencies for 108 stations in the eastern
and central United States in relation to lunar positions for from 10
to 20 year periods. Using epoch analyses, with the full moon as the
key day, she found a highly significant enhancement of thunderstorm
activity on the second days after full moons. The effect was even

greater when the full moon was near maximum north declination (closest to the geomagnetic tail). These reports are commensurate with the frequently published increase of maximum rainfall that occurs from 3 to 5 days after the full moon and, to some extent, after the new moon.

Potential Mechanisms of Lunar-Earth Weather
Relationships

Two general mechanisms exist by which the synodic lunar month could influence terrestrial processes: a direct effect from gravity variations and an indirect effect by blocking solar wind and/or disturbing processes in the geomagnetic tail. The first mechanism was attractive to early physicists since the center of mass between the earth and the moon lies about 1,000 miles below the earth's surface. The absolute position of this center changes in a circular manner every day and in a serpentine manner every lunar month. The essential argument posited a similar gravitylike effect on atmospheric processes.

Although the gravity theory is still important for "trigger explanations" of seismological events, recent satellite data favor the second hypothesis. The repeated occurrence of fast-acting processes around the new moon and the full moon and slower-acting processes shortly thereafter have been the primary supporting data. The new moon effects could be explained in terms of its "blocking position," since during this phase it passes between the sun (the solar wind source) and the earth.

The full moon effect requires the discovery of the geomagnetic tail and neutral sheet within which charged particles could be trapped. At the distance of the moon, about 60 earth radiuses, the width of the neutral sheet across the lunar orbit has been estimated to be 20 earth radiuses. Since the moon's orbital velocity is equivalent to about 14 earth radiuses per day, the moon would be within the neutral sheet for from one day before to one day after the full moon.

Synthesizing the assumptions of Fraser-Smith (1969) and Lethbridge (1970), the early passage of the moon through the geomagnetic (magnetospheric) tail could generate Hm emissions on days before the full moon. During the two days in the tail, particles trapped within the neutral sheet would receive sufficient perturbation energy to flow into the stratosphere, contributing statistically to the increased tropospheric nuclei and the increased thunderstorm or precipitation from two to five days after the full moon.

Lunar/Nonhuman Animal Correlations

Certainly the most reliable and frequent correlations between lunar phase and nonhuman animal behavior have involved invertebrates, especially worms. Massive amounts of data have been collected from the Palolo worm in the South Pacific to the monumental work of Frank Brown, Jr. and his colleagues in the laboratory. Laboratory rodents, shielded from the obvious cues of lunar light, an important factor for animals in the field, display similar but more complex lunar relationships.

If any species has been abused to prove the relationship between life forms and lunar cycle, the winner, by far, is the Palolo worm. This worm is reported to spawn only during the third phase of the moon in October and November in the South Pacific. Although frequently used as proof of lunar "driving," the worm's reproduction patterns prove little. In many of the early studies, no controls were used to determine whether the worm contains endogenous cycles that work independent of exposure to lunar phases or whether animals spawning during this time of the year and in that locality are more likely to survive environment and predators and, hence, to pass on their mating characteristics to their offspring. There is a great difference in the complexity of humans and worms, most of the time.

Frank Brown, Jr. has spent several decades determining the subtle factors that influence orientation angles of snails and flatworms. The tests are simple and replicable. Snails are maintained in a funnel-shaped corral so that only one can leave the entrance at any given time. Outside the entrance, an evenly illuminated area is divided into sectors; from far left (-7) to far right (+7), approximately 180° with the 0 sector directly in front of the chute. Brown found that the emergence angles of the snails show a consistent, although complex, variation with lunar phase.

Similar less complex patterns were noted for flatworms. The variable angle of emergence away from a light source, which flatworms avoid reflexively, changed from a -10° of center (0°) during new moon to +10° of center during full moon. The effect was most pronounced during winter months and when the emergence direction pointed north.

During spring, the orientation fluctuations became semimonthly and showed smaller amplitudes. Superimposition of a 4 gauss magnetic field could eliminate even the most intense lunar period effect (Brown, 1963). Several variables determine this periodicity. Apparently, the effect can be easily masked when a number of key variables are not held constant (for example, see Brown and Park, 1967).

The relationship between lunar periods and mammal activity is more difficult to demonstrate. Since light/dark cycles are powerful zeitgebers for rodent activity, test animals must be maintained in constant low-level illumination so that other zeitgebers can be effective. Without this procedure, any weak solar or lunar periodicities are masked by the daily light/dark cycle effects.

Frank Brown and his colleagues (Brown, 1965; Brown and Park, 1967) have published supporting data for a lunar periodicity in animals housed away from usual environmental cues. Over typical observation times of two years, weak but persistent increases in the activity of test hamsters occurred about five days after the full moon, with less intense peaks about the same time after new moon. With trough values for activity (a few days after the first quarter and the last quarter) of about 45 units, the peak values are about 55 units. The daily activity data contain much variability.

Lunar Phase Correlates of Human Behavior

Less than 100 years ago, published physicians still considered the full moon the central cause of psychotic disorders. Less than 50 years ago, some psychoanalysts had "isolated" a special kind of somnabulism as a consequence of the full moon. Today, throughout most of the countries in western culture, ambulance drivers, nurses, and police officers expect increased "weird" behaviors during the full moon period. With such a biased loading in the history of quasimedicine, it is not surprising that the majority of all lunar-human studies involve some measure of psychiatric behavior.

The first criterion in a controversial area such as lunar-human correlates should not be the "rational" nature of the effect, but rather the rigor of the methodology. About ten of twenty studies concerning lunar correlations and human behavior would never have demonstrated an effect even if one had been there. Types of analyses have ranged from indiscriminate clumping of data into either a waxing or waning phase of the moon (a 0, 1 condition) to actual day-by-day tabulation. Consequently, negative results as well as positive data should be closely scrutinized before a decision is made.

There is a tendency for reviewers of research areas to tabulate positive versus negative studies, as if the final sum determines the outcome of the phenomenon. This practice is very common in borderline areas. One can identify such "analyses" by all-or-none comments, such as, "The moon makes people mad." There are just too many complications and theoretical details that must be checked by rigorous methodology rather than cheap and simple publication rates. In this section, only samples of different methods are reported.

Weiskott (1974) found no lunar evidence in 736 counseling calls taken over a four-month period, but this decision was based upon a division of the data into either new moon or full moon partitions. Similarly, a study by Bauer and Hornick (1968), who recorded the numbers of people entering emergency service at a New York municipal psychiatric hospital on the days of the new, first quarter, full, or last quarter moon, showed no effect. On the other hand, Blackman and Catalina (1973), who checked emergency room admittance in a mental health center over a year period during 1971-1972, found a mean 30 percent increase on full moon days, a weak but statistically significant effect.

More agitated behaviors, specifically homicide and suicide, are reportedly full-moon dependent. In an almost infamous study by Lieber and Sherin (1972), homicides committed in Dade County, Florida between 1956 and 1970 showed small but significant peaks on full moon days and a few days after new moon. The trend was not reinforced when an Ohio county was used for the years 1958 to 1970. Peaks in homicide occurred a few days after the new moon and around the last quarter. In all instances, the peaks were very close to expected normal deviations considering the great day-to-day variance.

In a more recent study, Lieber (1979) correlated 11,613 aggravated assault cases over a 5-year period with lunar phase. The smoothed data, which is certainly more impressive than the legitimate raw data, demonstrated a clear peak around the first quarter and full moon, with lower peaks during first and last quarters. With mean values of 387, the full moon peak is associated with 25 assaults over mean values and the troughs are associated with 25 assaults under baseline. In short, the trough to peak effect from lunar phase is about 10 percent of the mean.

Lieber's work is inconclusive. On the one hand, there are periodicities reflective of lunar phase relation, though small in magnitude. On the other hand, different places show different types of periodicities. Although one can argue that lunar effects, such as tidal forces, should show different patterns in different localities as a function of latitude, comparisons of only two or three localities are little more than two or three points on a graph. In this instance, the locality factor becomes more of a rationalization than an explanation.

Suicide data are less systematic still. Lieber found a nonsignificant trend for more suicides a few days after new moon, at full moon, and at last quarter, but fewer suicides during first quarter and a few days after full moon. However, Ossenkopp and Ossenkopp (1973), who analyzed self-inflicted poisoning (for example, drug overdoses) in Winnipeg, Manitoba, Canada between February and August of 1971, showed a significant increase in suicides around first quarter, 40 percent over chance expectancy. A similar 40 percent increase of suicides over chance expectancy during the first lunar

quarter for subjects in Melbourne, Australia was reported by Taylor
and Diespecker (1972) for the years 1970 and 1971. In both studies,
the effect was only significant for females.

The third major traditional correlate with lunar phase is human
reproduction cycles, specifically menstruation and birth. Statistical
analysis of about 7,500 menstruations, over a year and involving
810 high school students, by Malek, Gleich, and Maly (1962) indicated
a tendency for menstruations to occur most frequently during the
first and last quarters and least frequently during new moon. The
effect was about 10 percent of the mean.

Menaker and Menaker (1959) determined the mean rate of births
for each day of the lunar month over a 10-year period. A total of
510,000 births in all the municipal hospitals in New York City and
10 private hospitals were used. Their data, expressed in percent
deviation from the mean, is shown in Figure 21. There is a clear
predominance for births to increase around the full moon and to
remain elevated relative to birth frequencies between new moon and
full moon.

Theories of Lunar Effects

Even when one removes the fancy nomenclature and parenthetical
scientific statements, lunar theories, in theme, are still reminiscent
of the "mysterious force" concepts that held popular appeal by superstition. The two most common forms of lunar explanations, outside
of outright rejection that any lunar periodicity occurs in human affairs,
are gravity and electromagnetism. Protagonists of a given theory
always assume a singular cause of the lunar relationship and ignore
the contributions of learning factors.

Bland statements, such as lunar-human correlations are direct
effects of gravity or electromagnetism, are from the same conceptual
core as mind, God, and hypnotism; their use as scientific explanations
are more a sham than a science. The most recent ardent theorist
supporting the gravity explanation has been Lieber (1978b), who argues
by analogy. He states that since the moon influences the tides and a
human being's volume is about 80 percent water by weight, that,
therefore, the moon must induce minitides within the human. These
minitides are responsible for the occasional displays of aberrant
behavior.

However, analogies are not scientific theories. If gravity is
indeed critical factor, this must be demonstrated in a quantitative
and systematic manner, after the obvious confounding factors have
been removed. The fact that any induced forces within the human
body would be extremely, extremely small is not at issue here. There

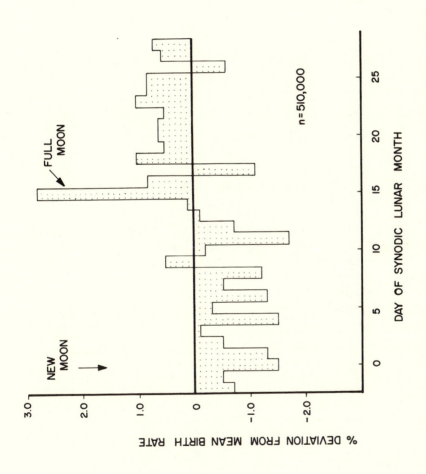

Figure 21 Distribution of relative birthrates as a function of lunar phase at the time of birth.

are many other biological phenomena discovered in recent years that occur at intensities or concentrations hereto prohibited by theory or expectation. One example is the extremely sensitive response of a part of the brain that contributes to water balance (the subfornical organ) to 10^{-15} mole (a fentomol) concentrations of the hormone angiotensin.

Lieber's lunar theory shows an absence of quantitative prediction and no capacity for systematic extrapolation. If gravity is involved, the behavioral effects should show variations as a function of lunar daily changes, perigee/apogee enhancements, and declination of lunar orbit with respect to earth. If subtle effects upon water within the human body volume are important, factors controling this important compound, such as antidiuretic hormones, angiotensin, and certain electrolytes, should vary systematically with lunar periodicities.

Systematic extrapolation based upon available detailed understanding of biological systems is a necessary exercise if lunar theories are to be removed from the amorphous-force explanations. For example, if fluid is the critical feature, one would expect significant changes in hydrodynamic forces within volumes typical of capillary space. Once would further expect the effect to be statistically enhanced within areas of the brain demonstrating high capillary densities.

Three of the most densely capillarized portions of the human brain are the paraventricular nucleus of the hypothalamus, the supraoptic nucleus of the hypothalamus, and the locus ceruleus. The paraventricular nucleus is associated with the release of oxytocin, which stimulates uterine muscle. Small bursts of this substance near term can initiate birth. The supraoptic nucleus is associated with the antidiuretic hormone vasopressin. The release of this substance increases the amount of water retained in the body by preventing loss through urination. This type of water retention is an important factor in menstrual swelling in the legs and abdomen. The third structure, the locus ceruleus, is a multifunctional structure best known for its control of dreaming and the sensations of fear and anxiety in human subjects.

Once specific areas of effect are isolated qualitatively, specific predictions could be checked. Does lunar phase, for example, contribute to the release of oxytocin in near term females, a factor that could contribute statistically to the full moon peak reported by the Menakers? If the area influencing vasopressin is influenced, quantitative changes in urinary volume and fluid consumption should be evident as well as swelling indexes in the susceptible areas. These factors should follow lunar periodicities.

Disruption of the normal activity of the locus ceruleus would reduce the relative amount of dream sleep at night and increase the probability of dream states in nondream periods of sleep or even in the waking state. Disruptions in dream sequences and sleep in general are well known antecedents in the later increase of irritability. Bouts of dreaminess frequently coincide with deviant behaviors during which the person doesn't "feel like himself." All of these factors must be examined quantitatively if any real support is to be given to direct physical lunar effects.

The second primary pattern in lunar theories is a type of electromagnetic field argument, a close conceptual cousin of gravity explanations, heavily loaded by the factor of invisibility. The works of L. Ravitz (1953) are among the most quantitative of these theories. Ravitz found that the overt behaviors, private experiences as recorded in the patients' reports, and d.c. potentials between head and chest varied with lunar phase. The majority of Ravitz's subject pool was taken from various schizophrenic categories.

Despite the methodological limits of the study, Ravitz repeated measurements between d.c. potentials, symptomology, and lunar phase. Precariously anecdotal at times, Ravitz reported, for example, a 34-year-old patient who showed d.c. potentials of +200 mV (about +2 mV would be normal) during the day of the new moon. These increases were associated with increased subjective experiences of dreaminess, as reported by the patient, and with increased objective signs of irritability and introversion. Lower potential readings were associated, for this patient, with increased feelings of alertness.

Ravitz instituted several internal checks of the crossvalidity of his observations. His first step was to determine whether the alterations in d.c. potentials within his population of schizophrenic patients could be used as a predictor for the general population. Assuming that increased psychiatric behaviors in the general population should be reflected in increased numbers of hospital admissions (a relatively safe assumption), he correlated admission rates with the day-to-day d.c. measurements of his patients. An unusually high correlation of 0.79 was found between the mean field intensifications of the group and the admission rates two days later.

Some of Ravitz's patients showed a semilunar peak in d.c. potentials and exacerbation of symptoms associated with the new and the full moon. However, not all subjects showed this effect; some showed lunar monthly variations with extremes, positive or negative polarities, at either the full or the new moon. The parsimony of the data is reduced when one examines the charts and graphs closely. There are not only clear noise problems (day-to-day variations), but also seasonal variations upon which the lunar month effects are superimposed.

The problem of individual difference, a central difficulty in biometeorology, is clearly evident in Ravitz's studies. This is an expected consequence of the gross and overinclusive nature of psychiatric categories such as schizophrenia. Considering an agreement rate of only about 50 percent between different psychiatrists or psychologists for the same patient and the multicontributory nature of schizophrenic behavior at the biochemical level, one can understand, in principle, the variability. In a functional sense, defrocked of the medical magic, the term "schizophrenia" is used in a manner similar to "weird" or "crazy." There are many, many distinctly different contributors to "crazy" behavior.

However, individual variation can also become an excuse for anecdotal proofs and absence of support data. Ravitz's data smack of this possibility; whereas one patient may show a d.c. potential polarity change as a function of lunar day, another will show the change every 14 days, whereas still another may show lunar effects only during certain measurement times. At the extreme, this type of reasoning becomes a type of change versus no change argument and the concerned biometeorologist is left with a binomial population, such as flipping heads and tails. In this instance, half of all the patients would appear to change in a nonspecified manner to some aspect of lunar phase.

Explanations of Lunar Effects

The demonstration of reliable but weak correlations between lunar phases and psychiatric admissions, suicide/homicide occurrences, and birth frequencies is an easy procedure compared with the isolation of mechanism. All of these behaviors have also been linked to geomagnetic disturbances, barometric pressure changes, weather instability (frontal movements), atmospheric ion contents, and thunderstorm activity. These environmental parameters have, in turn, been correlated repeatedly with the lunar cycle.

One of the few studies to consider weather parameters as well as lunar phase at the time of behavioral measurement was conducted by Geller and Shannon (1973) at the Lakeshore Psychiatric Hospital in Toronto, Ontario, Canada. They found, as have many other experimenters, that hospital admissions and patient contacts increased during full moon. However, during the measurement interval, the full moon was also associated with hot and humid weather.

Since hot and humid weather interferes with the normal sequences of sleep behavior and disruptions in sleep cycles have been strongly linked with deviant behavior the following day, one would expect an enhancement of psychotic or parapsychotic conditions. Such explanations would be somewhat more parsimonious and appealing than vague gravity or electromagnetic theories.

At present, the mechanism or mechanisms by which lunar periodicities are displayed in observed human behaviors cannot be determined. In essence, we are faced with the problem of a physician who is told that the patient has a runny nose and feels sick. With those two symptoms, the ailment could be a cold, an allergy, the flu, or the consequence of heroin addiction. There are not enough symptoms to differentiate the actual source variable.

The behaviors allegedly linked to lunar periods are also heavily influenced by psychological expectancy, incidental learning, temperature and humidity, geomagnetic disturbances, ion concentrations, and, perhaps, by as yet unspecified variables. Because of the vague nature of the behaviors, it is impossible to determine at present how much each factor contributes to lunar correlations.

Considering the weak nature of the effect, a statistical explanation appears the most valid and reliable. A statistical explanation indicates that each of these factors elicits small changes in susceptible members of a population. Some of the members may be more susceptible to some factors than to others. Individually, the consequences of these effects would be too small to appear above baseline variability. When all of these factors are presented within a short period of time, perhaps triggered or driven by the complex interactions between lunar phase and the terrestrial environment, the small effects are summed together. Summed together, they emerge above baseline values and, hence, become obvious to the observer. The full moon would contribute to the effect by serving as an obvious and convenient time reference.

From this model, the amplitude of lunar periodicities in human behavior is actually greater than it should be. These periodicities do not direct a strong, magnitude effect, but become apparent by contrast enhancement. The synchronization of the different small factors by lunar period would stimulate the unstable members of a population, who would have been distributed statistically and, consequently, would not have been obvious. When the unstable members of the population display their behaviors, the period after the burst is associated with a statistical decrease in available unstable members that might respond, thus increasing the net "amplitude" of the effect.

PLANETARY-SOLAR-TERRESTRIAL CORRELATIONS

The final level of analysis within the limits of this book involves the solar system as a unit. The term solar system, in a technical sense, implies a reciprocal interaction or relationship between the sun and planets. Although the sun is the major mass of the solar system, the plastic nature of its spherical characteristics and the periodic culmination of weak planetary forces, which are almost insignificant individually, indicate the existence of some distortion through presumably gravitational means.

The demonstration of the effects of planetary position upon solar activity has been a popular preoccupation among small groups of astrophysicists and quasiscientists alike. While quite possible in principle, practical demonstration has been plagued by the requirement of sophisticated mathematical techniques. Many of the epochs involved with planetary positions involve years or decades, a factor that reduces the number of analysis intervals to fewer than 20 (10 year epochs × 20 = 200 years).

The isolation of planetary effects upon solar activity from the background noise of multiple environmental factors is a mammoth methodological challenge. Not only are the epochs between certain feasible planetary positions time-extensive but, when they do occur, their effects are expected to be nonlinear. Unfortunately, sophisticated mathematical tools are often beyond the comprehension of the scientist who uses them.

Calculated values for the effects of tidal (gravitational) oscillations upon the sun have been published by several authors. One of the earliest candidates for tidal influence was the most massive planet, Jupiter, a speculation that was reinforced by the close similarity of the sunspot periodicity (about 11 years) with this planet's revolution period. Takahashi (1958) calculated other predominant periodicities: 2.89 months associated with the revolution of Mercury around the sun, 3.90 months associated with Jupiter-Venus positions, and 9.59 months associated with the synodic period of Venus and earth. Other relationships would be negligible according to his assumptions.

The Correlations

One of the more sophisticated mathematical analyses of planetary contributions to solar activity was reported by Bureau and Craine (1970). On the basis of relatively complex and assumptive analyses, they conclude that the combined effects of two or more planets contributed to solar rotation speeds (related to sunspot components). Various combinations of orbit times for Jupiter, Saturn, Uranus, and Neptune contributed to a double period sunspot cycle: 22.46 years.

Wood (1972) has argued that the tidal influence of planets upon the sun depends upon their size and distance. According to Wood, there are only four planets that would display appreciable effects upon solar tides. Taking the earth's contribution equal to 1.00, the relative contributions of each of the others is Mercury 1.15, Venus 2.17, and Jupiter 2.28. The remaining five planets would make only negligible contributions. However, nonlinear combinations between the distant planets and Jupiter, as calculated by Bureau and Craine, are not assumed.

Presumably, when Venus and earth are in line (conjunction or opposition), they cause a solar tide 50 percent greater than the changes induced by Jupiter. Tidal height fluctuations associated with earth-Venus conjunctions and earth-Venus oppositions are highly correlated with peaks in sunspot numbers. The data are impressive and consistent between 1750 and 1975. His forecasts for the next sunspot maximums, based upon previous sunspot cycles, are 1981-1983 and 1992-1994.

The Nelson Report

Correlations between planetary positions and ionosopheric activities are expected as secondary consequences of planetary tidal effects on solar activity. Surprisingly, few correlations have been reported for this relationship. The most notorious and too frequently quoted, however, is a questionable analysis by Nelson (1951). Since this study has been quoted by scientist and pseudoscientist alike to prove everything from complex astrophysics to simple astrology, it deserves special attention.

Nelson's procedure was simple. Dates on which the heliocentric (sun center) relationship of any two planets was 0° (a time when an inner planet was in line on the same side of the sun with an outer planet), 90°, 180°, or 270° were called configurations. Every configuration was calculated between Mercury and Venus, Mercury and earth, and so on, out to Saturn. Similar configurations were calculated for the other planets in a successive manner. For reasons that are not given, Nelson selected only five years of an eight-year period for analyses.

The measurement of ionospheric disturbance was based upon radio technicians' evaluations of whether a given eight-hour period was "poor" or "good" for radio communication. Each day contained three such ratings. No data were given with respect to whether the ratings selected were averages from more than one log or whether they were selected after the configurational dates had been obtained. No elementary but essential data on the distribution characteristics of the measurement (an all-or-none condition) were given.

Nelson's data appear impressive. Graphs for each of the five years indicate disturbed periods to be from three to five times more likely on days of some configuration compared with from three days to four days before or after the configuration. The total number of disturbed periods per year ranged from about 125 to 200.

However, the study, for the most part, has little scientific value. Essential methodological details and fundamental statistical practices are absent. The reader is not told how many configurations (and, by inference, there must have been hundreds) occurred during the analysis interval. Such data are important to determine the likelihood by chance alone that the variations of poor versus good would occur.

Nelson uses the term "disturbed periods" as his dependent measure and days (for example, before and after a configuration) to present the effect. The reader is not told how he selected periods on days when all three periods displayed poor ratings. From the style of presentation, it appears that Nelson summed all disturbed periods in a given day. If this were done, he loaded his effect by at least a factor of two. This procedure is not only statistically invalid, if not tantamount to fudging data, but would also have exaggerated any weak effect.

The only other data presentation is anecdotal and fragmented. Sample diagrams are given of planetary positions on particular disturbed days. There is no indication of the tolerance involved with the "right angle" effect. Are the alleged increases in disturbances still evident with configurations of 92° instead of 90°? More importantly, to what extent did Nelson accept configurations that were not quiet right angles, especially if they occurred near a poor transmission day?

Nelson repeatedly used the term "correlated" to indicate a relationship between planetary configurations and poor transmission days. It is clear, much to the disappointment of the serious reader, that this term is used in a subjective manner. Neither a single correlation coefficient nor quantitative measurement of relationship is presented in the paper.

Nelson's paper is an unfortunate event. In a review of borderline and pulp book publications, I have seen this work referenced more than 100 times. Close scrutiny of the authors' comments clearly indicate they have not read the article or at least have not understood it. Indeed, there may be such an effect on ionospheric disturbances—our reliable understanding of weak planetary-solar-ionospheric interactions demands this. However, Nelson's paper is so full of methodological flaws that nothing concrete can be concluded.

Direct Planetary Modulation of Terrestrial Phenomena

Whereas indirect influence of planetary position could influence weather through alterations of solar activity, Mercury and Venus are in a special condition to influence the earth directly. Since they are between the sun and the earth, they could influence the flow of solar wind and hence influence the occurrence of geomagnetic disturbances.

Strong evidence exists for this possibility, according to the data of Bigg (1963) and the analyses by Jacobs and Atkinson (1967). Bigg found that numbers of geomagnetic disturbances decrease from an average of 50 over the analyses interval to about 10 during periods of inferior conjunction between earth and Venus (that is, Venus between the earth and the sun). The effect was evident for the inferior conjunction angle plus or minus 10°. However, after Venus passed between the earth and the sun, the number of geomagnetic disturbances increased by a factor of two (~100) for about a 10° sector before returning to baseline. Inferior conjunction between earth and Mercury had less intense and more complex effects.

Takahashi (1958), who calculated the synodic values for earth-Mercury and earth-Venus conjunctions and solar effects, contends that a relationship exists between the position of these planets and changes in barometric pressures. Using anomalies (variations) in the height of the 500 mb pressure levels, he reports alterations sharply associated with these positions. Again, the data are presented in an unusual manner. Interpretation and evaluation are difficult.

The existence of earth-Venus and earth-Mercury interactions are not surprising in light of the demonstrated alteration of solar expansion energies by lunar phase (in this case, the new moon). However, the actual geometry of alteration appears questionable in light of the small diameter of these planets in context of their distances. Models vary from direct gravitational influences, to blatant shielding of solar particles, to the consequences of a turbulent wake within the fluidlike conditions of the solar wind.

Bioplanetary Correlations

Few reliable or valid correlational studies between behavioral measurements and planetary positions have been published. The most obvious confounding variable for these studies is the many solar perturbations that would be involved with planetary positions. Such factors would modulate any enthusiasm about the importance of the effect and completely eliminate, for now, the isolation of mechanism. Since the intensity of the behavioral effects would no doubt be very small (even with very, very liberal assumptions) and require large samples (in the order of thousands of subjects) for demonstration, internal quantitative predictions as means of isolating various mechanisms would not be possible.

A relatively well-designed study was published by Randall and his colleagues (for instance, Rogers and Randall, 1972), who observed changes in cat behavior and physiology. Initially, they were studying the variations in thyroid activity and various metabolic measures in cats as a function of sunshine hours. However, they also found a clear three- to four-month rhythm in cat thyroid activity in both their North Carolina and their Iowa laboratories. The rhythm appeared independent of sunshine hours, temperature, barometric pressure, or relative humidity.

In the search of a possible zeitgeber for this relationship, Randall and his associates noted that the synodic period of the planet Mercury was 115 days or from about 3 to 4 cycles per year. Dividing the angle between lines drawn from earth to the sun and from earth to Mercury by the square of the distance between earth and mercury to obtain a quantitative measure of variation, they found a correlation of -0.49 for the Iowa cats and -0.67 for the North Carolina cats between this measure and thyroid activity.

The Randall correlations only demonstrate that planetary positions, especially the short-term periods as Mercury or Venus, can be demonstrated in biological oscillations. Although mechanisms at the astrophysical level are available for these two planets in particular, the means by which the planets could influence biological systems via weather variations is unknown, assuming the relationship really exists. There are a vast number of overlapping cycles in the millions of species studied and thousands of weather parameters measured. By chance alone, one would expect a few persistent overlaps.

11

PERSPECTIVE
AND POSSIBILITIES

Ultimately, the journalist, the politician, and even the scientist prefers to have a simple, all-or-none answer to the question: Does weather significantly affect human beings? (Implicitly: Are there any large, as yet undetected effects?) Some require such an answer for brief discussions with people who have little scientific training. Others demand a simple answer as the personal and emotional solution to an apparently insoluble problem.

Unfortunately, from the perspective of brevity, all questions are not created equal. Some questions have simple yes and no answers, while others have qualified answers. Many questions are not applicable to the situation. There are a few questions that are nonsense; they have no answers at all.

The biometeorology question occupies a variable position between the extremes. If weather were a nominal word, that is, all or either of one thing or another, there would be little difficulty. If each human being were a singular, homogeneous set, the consequences would be clear and simple.

If weather were like the terms U.S. citizen or dead versus alive, there would be little ambiguity. Either a person is or is not a U.S. citizen. A person is either dead or alive. There are no intermediate options. There are no conditions of being 45 percent of a U.S. citizen or 67.5 percent dead.

In some instances, a person may show simple and discrete behaviors. If a person must vote Republican, Democrat, or other, the choice is simply one out of three. Implicitly, this assumes that the person is homogeneous: he only votes and displays no other behaviors.

However, the human being is a very complex system. Some of these response systems are affected by stimuli from the social and physical environment, while others display little activity. As mentioned in Chapter 3, the human being can be seen as a complex ensemble of response systems. The nonchalant mixing and combination of these separate systems only forms a mishmash of confusion and insolubility.

So, does weather affect human beings? There is no simple yes and no answer. Since weather is composed of a variety of different bioeffective stimuli presented at different times and at different intensities, one must consider each of these conditions. Since the human being is a complex set of varying response systems, influenced significantly by learning history, simple extrapolation to the entire population is not reliable.

The serious biometeorologist must be content with those fragments of facts that have been verified by careful experimental or correlational analyses about particular behaviors or disorders associated with specific aspects of the weather matrix. Two questions should be asked repeatedly: How much of the variance in that particular measure of human behavior can be accounted for by the weather? Are there any obvious or suspected confounding factors?

GENERAL SUMMARY

The most frequent component of the weather matrix to be correlated with human behavior has been temperature. Unfortunately, the exact degree to which temperature influences weather-related complaints has been masked by the common procedure of data pooling. Many writers have lumped daily or weekly variations in temperature together with seasonal or monthly variations. As a consequence, the problems of transient stimuli (weather-related temperature changes) have been confounded with baseline conditions (seasonal-related temperature changes).

Temperature

The distinction between short-term air mass-related changes in temperature and seasonally related changes in temperature is more than just an academic exercise. To explore the effects of weather upon human beings, the differentiation between baseline or steady state conditions and transient variations (weather) superimposed upon the baseline must be maintained. The dangers of homogeneous labeling have been discussed earlier.

Although the avid experimenter is prone to select one component of the weather matrix—usually one of the more exotic components, such as barometric pressure, magnetic fields, or ions—as the important stimulus of weather effects, an objective evaluation indicates that the normal variation of these components is trivial in comparison with temperature. Variations in air temperature always appear to precede or accompany the more popular component of the weather matrix.

The effects of temperature upon other components of the weather matrix are almost impossible to partial out or to control. Temperature variations, first and second derivatives, have direct and indirect consequences on both primary and secondary components of the weather matrix. Even a simple factor analysis indicates a first-order common factor upon which temperature variables are highly loaded.

Direct effects of temperature are evident within the concept of an air mass. Thermal energy within an air mass determines the distance between air molecules and, hence, can influence the barometric pressure. Thermal potential within an air mass also determines the solubility characteristics of the medium. The amount of water vapor that can be maintained within a parcel of air without condensing varies proportionately with temperature.

Often, even with simple correlational studies, the contributions of temperature variation are obliterated or masked by crude or simplistic analyses. Frequently the relationship between these variables and temperature may have intervals of nonlinearity. Any correlation calculated with these data would not demonstrate a significant relationship between temperature and the other measure.

Secondary consequences of temperature variation are more difficult to control. Even within the comforts of a house, in which temperature is controlled by thermostatic mechanisms, the effects of external temperature cannot be ignored. Ultimately, the air one breathes comes from the outside. During the winter, the air temperature must be elevated to comfortable levels by artificial heating, a procedure that depletes the relative moisture content.

Rehydration of the air by humidifiers attenuates the problem somewhat. However there are still other indirect consequences, such as greater production of air ions during winter periods. Since most heating sources involve the heating of metal parts (wallboard heaters, air ducts), large numbers of positive ions can be generated. Artificial hydration to normal levels may not reduce the concentration of these ions to summertime levels.

Changes in external temperature induce a variety of time-varying phenomena that can influence human behavior and diseases. Thermal gradients across walls or windows can generate small breezes that can slowly but significantly remove the heat from body surfaces. These processes are antagonized by wind.

Behavior

The types of normal behaviors and disorders associated with
the weather matrix have clear temperature-related mechanisms.
Since the peripheral circulation is one of the first systems to respond
to small changes in external temperature, it is not surprising that
various peripheral circulatory disorders have been correlated with
the weather matrix. Other disorders, such as glaucoma, arthritis,
and general stiffness, are related to skin temperature and, hence,
to vascular flow.

Severe alterations in peripheral blood flow, such as periods
of sudden heat or cold, place greater emphasis on the central vascular
system. The consequent dilation of the heart vasculature during cold
stimulation or constriction during hot stimulation can lead to failure
of already damaged or unhealthy tissue within the heart muscle. The
various forms of heart failure have been discussed.

With such emphasis on vasculature, one would predict that
weather fluctuations are correlated with various categories of stroke,
especially in elderly patients. Data must still be collected to determine whether weather-induced alterations in brain blood flow, which
is influenced both by chemical (noradrenalin) and neurogenic (acetylcholine containing terminals around certain arterioles) sources, can
influence deleteriously the vascular rich portions of the brain. Such
areas would involve, interestingly enough, those centers associated
with water balance and thirst, oxytocin release, and dreaming.

Other behaviors associated with weather are known also to be
correlated with temperature. Traffic accidents, bouts of aggression,
sexual assaults, incidence of suicides, and general agitation are
conditions known to increase statistically during temperature extremes. In these instances, temperature is certainly more of a
trigger than the primary cause, although some theorists have argued
that the spontaneity of many antisocial acts suggests a strong, local,
and transient stimulus outside the offender.

The second class of behaviors appear to be those best labeled,
for now, as the thyroid activity cluster. They range from general
hypoactivity to specific symptoms with schizophrenic-form complexity.
Such behaviors have longer response latencies and are displayed for
longer periods of time. Frequently, they involve slow baseline
changes in overall organismic activity that may not be detected by
the individual.

The most obvious source variable of the thyroid cluster is
basal metabolic rate. In addition to pernicious anemia, secondary
consequences associated with alterations in thyroid activity include
changes in the cellularity of bone marrow, gamma globulins and
modifications of autoimmune responses, and spleen or thymus activity.

Elevated thyroxine levels are also known to increase pentose shunt enzyme activity in red blood cells and, hence, oxyhemoglobin levels in the blood, but to decrease the activity of several clotting factors. Many of these effects may require from days to weeks to be evident.

The relationship between the thyrotropin releasing factor from the hypothalamus, the thyroid stimulating hormone in the pituitary, and the release of thyroxine from the thyroid is well known. This hypothalamic involvement and the recently established coupling between autonomic nervous activity and thyroid function allow the contribution of important psychological factors, especially those associated with classical conditioning.

The relationship between stimuli of the weather matrix and the thyroid is evident when one considers the intimate relationship between temperature and thyroid activity. Lesser known consequences of decreased thyroid activity include a variety of periodic disorders with 28-day and between 3 and 4 month durations. These include ulceration, arthritic complaints, insomnia, dipsomania, depression, and hypochondria.

Selective Population

To any stimulus or set of stimuli, one would expect a more or less normal distribution of a response measure. Weather-sensitive people do appear to exist, although the constantly confounding effects of expectancy and learning have never been removed even in the most well-controlled studies.

However, those people who are most responsive to weather changes appear to be those most sensitive to transient disturbances in the autonomic nervous system. Older people, who, as a population, appear to lose their ability to subjectively determine variations in their own body temperatures, are more responsive to weather changes. These individuals may, in extreme situations, actually develop hypothermia or hyperthermia without subjectively realizing the intensity of the distress.

Disturbances in vascularity and the highly correlated alterations in water balance often worsen preexisting disorders of these systems. The general failure of this population to respond and adjust to transient stimulation results in greater complaints of arthritis, general pain, lethargy, and stiffness. The secondary consequences of these conditions, such as greater likelihood of falling, slipping, or overstretching muscles, is evident as well.

One serious possibility seems to be unrecognized. Geriatric members of the population are susceptible to recondite necrosis that accompanies the transient failures in peripheral circulation. In domiciles in which homeostatic heating is not trustworthy (which appears to occur in the majority of low economic conditions even in the Western world), short periods of temperature variations around the mean may antagonize blood insufficiencies to already fragile tissues.

Although, no doubt, the blood flow ultimately returns to normal when the system adjusts (otherwise clear complications would arise), the cumulative effects of such low-level tissue damage would antagonize the connective tissue and possibly the immune system. It is not clear whether occasional exposure to the outside environment would alter this condition.

The second major population affected by weather changes appears to be some components of the so-called psychiatric population. Their sensitivity appears related more to the unstable nature of the autonomic nervous system rather than to its failure to respond normally. Unfortunately, a system that is very sensitive is frequently very unstable.

If one considers the experience at any given increment of consciousness to be a combination of weighted inputs from external sensory, internal sensory, and stored (memory) sources, the condition of the weather-sensitive person within this population can be realized. Under normal conditions, we either learn or have the genetic potential to repress "vegetative" inputs from the lower brain centers. Even during "private" or ruminating periods, most of our information is derived from the external or exteroceptive environment rather than from interoceptive (proprioceptive) cues.

To register a visceral input within "experience," most inputs must be relatively intense or protracted. With the exception of short or sharp pain, most visceral changes appear only statistically within the person's consciousness as the stimulation increases. The range in different people's recognition of flulike feelings to the presumably same viral stimulus level is quite amazing.

The sensitive person could be viewed as an individual in whom the balance is loaded toward the visceral end of the continuum. Small inputs from this or that organ, ignored by a person who has learned to ignore or who has never learned to label small visceral inputs, would be registered with greater reliability and distinction. Visceral inputs for these people would be weighed to such a degree that their inclusion into conscious experience would be a frequent and periodic condition.

Depending upon the person's labeling history (how they learned to label their internal states), low-level and, for the most part, nonpathological changes would be reported. A small vasodilation of brain vessels might be registered as an intense headache. Enhanced water retention would be experienced as a bloated, lethargic condition associated with increased irritability.

The most frequent response to weather stimuli is usually a generalized, low-level discomfort and restlessness. When pain labels are used, low-level throbbing pain is reported more frequently than sharp pain or localized pain. In some instances, the problem organ may appear as the focus of the pain, although the sensations are again vague and generalized. The reality of the symptoms is clear to the patient.

People who are more vagotonic (parasympathetic) than sympathetic dominant appear to display more of these visceral symptoms. Whether these patients have learned to attend to the increased input from the viscera or whether they passively respond to the greater number of visceral sensations characteristic of parasympathetic activity is not clear. Within an external environment of visual and auditory redundancy and the general boredom of ritual, the novelty of the visceral changes associated with meteorological conditions would have significant impact.

People who live in relative social deprivation (many older people), perceptual deprivation (invalids or people with limited transport), or great redundancy (homemaker neurosis) appear to be more responsive to novel but transient inputs from viscera. Their lives may be never ending trains of aches and pains that appear to move with unusual versatility in and out of different medical diagnostic categories. Not surprisingly, the family physician is usually incredulous.

Such weather-sensitive people have commonly appeared under the clinical categories of hypochondria, neurosis, and hysteria. Long convinced of the reliable nature of their experiences and of the fact that other people do not appear to respond to the same conditions, they are prone to be more self-centered and unresponsive to many social cues.

The primary differentiator between weather-sensitive people and the general class of borderline psychiatric conditions is the former's objective responsiveness to weather changes. These individuals display repeated correlations between their complaints and weather changes. Unfortunately, the literature does not demonstrate any reliable a priori means of determining the weather-sensitive person. Careful collection and cautious interpretation of the data are required.

HAVE WE MEASURED THE APPROPRIATE BEHAVIORS?

Even a liberal conclusion of weather-related changes in the most sensitive behaviors indicates that alterations in the weather

matrix or any of its components account for not more than 30 percent
of the variability in these behaviors. A more conservative estimate
is about 10 percent, even in weather-sensitive populations. At these
magnitudes, weather effects may be "a great ado about nothing."

On the other hand, biometeorologists may have traded off
behaviors of maximum weather sensitivity for ease of measurement.
Traditionally, they have measured numbers of accidents, deaths,
suicides, homicides, and other all-or-none type measures. These
behaviors are relatively drastic responses. They may not reflect
the everyday variations influenced by the weather matrix.

Everyday behaviors allegedly associated with weather changes
are diffuse and relatively small alterations. General lethargy or
excitement, sensations of well-being or depression, ability to concentrate, and simple coordination are more common responses.
They may involve periods during which the person stumbles while
walking, slams his knuckles against the wall when putting on some
clothes, or displays greater difficulty in understanding even the
most simple discrimination.

Another behavior prone to daily or weekly variation is fluid
thinking, that is, ease in chaining diverse thoughts. Peak periods
are followed by intervals of staring into space. Some people report
intervals of 2 days ± 1 day within which they cannot perform complex
subtle motor behaviors, such as typing, painting, or playing the
piano. During these periods, they report a feeling of "not being with
it" or of being "out of touch" or some other metaphor for low-level
feelings of loss of control.

These behaviors have not been measured directly or for sufficient periods to allow reliable analyses. No doubt these behaviors
would be implied by their consequences, such as variations in numbers of accidents. One would expect that such behaviors would be
evident in complex tasks, such as driving a car during rush hour
or operating a complex piece of military or industrial machinery.
However, these effects would be secondary.

Most of these diffuse and statistical behaviors are associated
with mood. The term mood is usually applied to protracted changes
in subjective evaluations of emotion. Mood is considered to be the
baseline upon which more transient bursts of emotions, such as
anger or fear, are superimposed. Crude mood measures indicate
that general autonomic feelings, that is, positive or negative affective
ratings, are lowest on Mondays and during January, February, and
March.

Relatively simple numerical scales (from 0 to 10) associated
with verbal descriptors (adjectives) demonstrate surprisingly reliable
relationships with weather. Several studies by this author suggest
that normal adult male and female subjects, who rated their moods

four times daily for several weeks, were responsive to weather alterations. As much as 50 percent of the variability in the mood rating could be accounted for by weather [very similar to the values reported by Goldstein (1972)].

The bulk of the explained variance was due to the temperature and geomagnetic conditions of the previous two days. A significant portion of the remaining variance in mood rating responded to examination schedules, local university celebrations (winter carnival), and episodes of sickness. All analyses involved daily intervals and did not consider weather phases or greater time increments. Synergisms between weather and social factors were not investigated.

THE REQUIREMENT FOR A RESPONSE MATRIX ANALYSIS

The acute thinker quickly realizes that the autonomic tone explanation verges upon the problem of the nebulous theory. If weather changes, alterations in reinforcement schedules or lifestyles, and daily uncertainties all influence autonomic tone, how does one differentiate between them?

The situation is similar to the historical difficulties involving the measurement of emotions. With a single measure, such as heart rate, sweating, or mood evaluation, one is left with a situation in which a given emotion either influences or does not influence the measure. Due to limited measurement, emotions that should be differentiated are combined.

Recent studies indicate that different emotional stimuli can be differentiated if there are sufficiently different, low-correlated responses within the analyses. If one measures a variety of blood chemicals, endocrine responses, operant responses, and routine vital signs, situations that were initially considered the same can be clearly discriminated. There exists a particular profile for different emotions.

By using a response matrix containing many responses from different noncorrelated systems, the experimenter has more degrees of freedom with which to make fine differentiations between stimuli. Even nominal measurements, that is, increase or decrease, can be used if there are sufficient numbers of measurements. With a large number of responses, one can determine the profile or the matrix of 0, 1 responses.

Suppose the biometeorologist had sufficient funding to measure, for several months, 15 different blood constituents (serum enzymes, electrolytes, and globulins), 10 different pharmacological responses (including electrocardiogram and electroencephalogram changes), and 10 different behavioral-medical measures (such as ambulatory and self-rating measures). Weather Type I could be associated with increases in responses (for example) 8, 9, 23, and 33, while Weather Type II may be associated with increases in responses 12 and 23 but with decreases in 1, 3, and 33.

With 35 different 0, 1 options, one could build a variety of different matrixes to stimuli that had been too diffuse or overlapping to be determined. Potentially, the organismic changes associated with taking vacations or changing jobs or getting bored with winter could be differentiated from the effects of various forms of the weather matrix.

PSYCHOEPIDEMIOLOGY AND WEATHER

The 1980's will be a period in which global reconnaissance of weather and of social dynamics will become a sophisticated science. With the advent of weather satellites, effective news coverage, and the gradual centralization of medical, social, and political data, information will be available within temporal intervals applicable to biometeorology. As daily recording of disease counts, antisocial behaviors, and related changes slowly replaces monthly and yearly averages, the relevance of the weather matrix will become evident.

Greater usage of computer entrance and storage of information will allow details about locality and (daily) time for a variety of weather-related disorders. Biometeorology will be able to progress past the first-order observations, which have been so crude that they are almost common sensical, to more subtle n-order observations. Within such a rich data array, the power of psychological variables will be resolved.

Conceptually, psychoepidemiology would be the study of the sudden and transient alterations in human behaviors that are highly correlated with traditionally private processes, such as thoughts or attitudes. These behaviors would involve widespread or epidemic occurrences of paranoia, irritable aggression, apprehension or anxiety, or general emotional disruptions. In more particular instances, psychoepidemiological responses could include bursts or rashes of purchasing (novelty objects), investments (such as wildfire land investments), or crowd violence in otherwise routine settings.

Frequently, these behaviors would be restricted to an area associated with similar meteorological and social stimuli. On the basis of general observation, one would expect the typical areas to involve air mass dimensions. More localized and intense effects could occur within areas 100 km^2, especially in those regions occupied for sufficient durations by unusual inversions or stationary fronts.

To date, the weather-related characteristics of many psycho-epidemiological behaviors have not been noted systematically. Although local government officials have acknowledged the epidemic nature of many psychological conditions during various obvious weather matrixes, studies involving larger areas have not been completed. Large scale application has been impossible until recently since air masses move quickly in space and time relative to the intervals in which behavioral data have been typically collected.

The characteristics of a psychoepidemiological response would be similar to other cascading processes. One would expect that the more sensitive portions of the population would be influenced first by the weather matrix. Later, other members of the population would begin to show various behavioral alterations.

A critical factor that has not been determined, conceptually, is the percentage of population that must be influenced psychologically by the weather matrix before significant alteration occurs in the entire population within the set. If the relationship between the relative percentage of the people affected by some weather matrix and the time required to produce this effect is linear, one would expect total inconsequential populational changes. Alterations from a nonlinear relationship may have more serious consequences.

Figure 22 displays some of the theoretical possibilities. Situation A would be relatively safe; it would not effectively influence the stability of the total population since the time required to change or affect a significant percentage of the population would surpass the duration of the weather stimulus.

Situation B would involve a nonlinear change that would be cascading rather than monotonic. In a geometric progression, more and more elements of the population would be affected by the weather change. However, at some point, the sensitive portions of the population would be totally affected and hence the effect would level off (curve B) or return to the original baseline (curve B^1). The effect would be constrained.

The serious contingency is curve C in which an all-or-none situation exists. Once a critical percentage of the population displays the behavior, the remaining portions of the population are influenced in an exponential manner. More and more components of the population show the behavior until the population is totally exhausted or saturated.

A very local example of type C response would be the chain-letter craze that periodically strikes university campuses. Within a given population of about 10,000 students, a single person introduces the concept of a chain letter in which the person buys a letter for a given amount. In order to make a profit, the student must sell it to another person. Each time the other person sells the letter again, the previous "owners" collect a portion of the money.

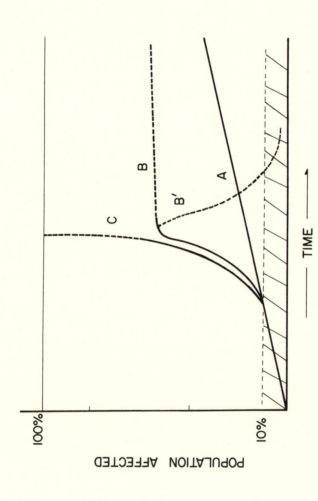

Figure 22   Various relationships between the percentage of the total population affected by a weather-
induced change in a small portion of the population (in this case, 10 percent of the population)
and the rate at which a response is displayed.   Condition A is a simple linear effect over time.
Condition B involves a response that is nonlinear but reaches a limit before asymptoting or
returning to baseline.   Curve C involves a response whose boundary is not predictable except for
the maximum value.

Initially, there is very little sensitivity to the situation. Most students reject the idea or consider it illegal. However, if a criterion number of students show the behavior—under certain situations— recruitment occurs in an exponential manner. Within 24 hours, the entire campus can be saturated. After this episode, the behavior is no longer displayed for months or years.

The bases of the psychoepidemiological phenomena would be the clusters of emotional responses influenced by the weather matrix. As Davitz (1969) indicates, there appear to be 12 discriminable clusters of emotional or autonomic conditions. Each one is associated with particular subjective and physiological patterns.

For example, one cluster is associated with "moving toward." Descriptors associated with this condition are "wanting to help," "realization that someone is more important than I am," "wanting to make others happy," and "I want to touch and be physically close to other people." Such conditions would facilitate greater interaction and maintained proximity to other people. They would facilitate also the spread of contagious viruses, if present.

Another cluster, associated with "moving against," shows another series of descriptors. They include, "I want to strike out," "I want to say something nasty, to hurt someone," and "I keep thinking of revenge." This autonomic tone would facilitate aggression and all of its manifestations from autoaggressions to homicides.

The cluster of hypoactivation is associated with psychological experiences such as feeling empty, drained, tired, and sleepy. A person experiencing this cluster would describe her psychological state as "I feel let down," "I feel mentally dull," or "I feel sluggish." Behaviors most highly correlated with the degree of alertness would be affected adversely.

More serious personal consequences in depressed patients could be fostered by either the discomfort or the inadequacy cluster. The former situation is associated with reports such as, "There is a heavy feeling in my stomach," "There is a sense of loss, of deprivation," "I can't smile or laugh," and "I feel as if I am under a heavy burden." The latter situation is associated with reports such as, "I feel vulnerable and totally helpless" as well as a general inability to cope with everything. A sense of "not knowing what to do" predominates.

Other relevant clusters would involve "enhancement" and "incompetence." Enhancement is associated with "a feeling of being taller and a sense of sureness." Muscle tone is suddenly increased while a sense of accomplishment, fulfillment, and higher intellectual ability predominate. Incompetence, on the other hand, is associated with the feeling that "nothing I do is right." In this cluster, the person desires a change. There is a yearning to have things as before and a sense of weakness and regret.

The rich variety of combinations within the weather matrix would allow the stimulation of each of these twelve known clusters of affective experiences. When a critical percentage of the population is experiencing one of these psychological conditions at the same time and within the same proximity, quantal changes in overt behaviors can occur. The effects would be strong but transient.

THE TENUOUS AND THE UBIQUITOUS

Phenomena that can "only be seen by the whole" have been suspect throughout the history of science. Homogeneous, amorphous "things" or "forces," such as orgone force, odyl fields, and mesmerism, have permeated the early stages of many sciences. Their existence is argued indirectly and by negation; they are, supposedly, what is left when everything else is removed.

Such elusive concepts are a consequence of semantic artifacts, crude measurement procedures, and strong beliefs. They are antagonistic to scientific behaviors. Sometimes these concepts appear as various forms of a god belief or as more sophisticated excuses for statistical (random) variability, such as "the hidden variable." Punctuated by such statements as, "You can't prove they are not there," they are powerful sources of confusion for the unwary thinker.

The word game characteristic of these concepts is the first sign of their true identity. Typically, they occur when the numbers of observations and different measurement procedures for a phenomenon far exceed the human observer's capability to understand. In these instances, the experimenter—still a human verbal animal—frequently concludes frustrating situations with statements such as "There must be something more," or, "There must be a ghost in the human machine." More frequently, the conclusion is a crude form of vitalism or some indestructible, omnipresent life force.

Biometeorology is an area extremely prone to tenuous and ubiquitous concepts for a variety of reasons. First, the data are not only endless, they are also complex and confusing; consequently, they appear insoluble. This factor alone makes biometeorology less attractive to the average scientist, who prefers to reduce uncertainty and anxiety rather than to intensify it.

Secondly, the tools of analyses foster a field mentality. Before the routine use of computers and statistics, biometeorological concepts were frequently reinforced by looking at complex graphs or figures. General trends were noticed without control for source variables to which many apparently unrelated variables are correlated or for common sources of measurement error.

The researcher faces a similar problem with computers and complex statistical methods, but for different reasons. Since multiple regression or a related statistical procedure is the most frequent model of analysis, artifacts from multicolinearity among the many independent variables from the weather matrix within the equation can elevate the multiple correlation coefficient in a spurious manner. This problem is of critical importance when one is dealing with low-level effects that explain about 10 percent of the variability in the response measured.

In this instance, suppose that temperature changes are correlated approximately 0.33 with some behavioral measurement. Such a correlation would indicate that temperature changes can account for only about 10 percent of the variability in the behavioral measure. Now, if humidity, barometric pressure, and wind speed were correlated 0.9 or more with temperature and all four variables were in the equation, one might get a multiple r-value much greater than 0.3. Since the independent variables are so closely intercorrelated, one could be merely adding the same source of variance four times to produce a very inflated and artificial effect.

On a more personal level, the "gestalt" or field mentality is almost impossible to extinguish. Biases of perception, the selectivity of memory, and the general poor observational skills of the human being foster a field theory. Within the strong influence of psychological closure, oddities or deviations from the whole picture are dismissed by a variety of rationalizations.

Biometeorology has had several episodes of the amorphous theory. When singular components of the weather matrix, such as ozone, could not explain all the behaviors measured, rationalizations, such as "ozone-like" substance, were made. When components of the weather matrix were carefully catalogued and the "weather substance" was never found, the physical characteristics of that substance acquired more and more magical characteristics. A variety of names have been given to that "substance," such as aran.

One can appreciate the easy predilection to evoke a magic substance as the substrate to explain the residual when the major weather matrix has been removed. After all, there are well known instances of synergism in pharmacology whereby the combination of two substances produces changes that are different or greater than the simple summation of the two substances. There are gestalt principles in perception whereby the experience of a square is independent of the objects that make the four 90° angles.

However, these theories are dangerous in biometeorology. They open a condition in which hypotheses are not systematically tested against data. If one assumes that a magic substance exists, hidden within the complex patterns of either the phenomena or the analyses procedure, then it can never be directly tested. There will always be ample numbers of anecdotal accounts that will reaffirm the belief.

If there is a common condition or complex interaction between weather stimuli that is primarily responsible for the more bizarre aspects of weather-related behaviors, then it should be detectable in some systematic fashion. Its presence should appear as a function of the number of variables in the analyses (e.g., the number of independent variables in the multiple regression) or of their intensity. In other words, it should systematically vary as a function of some combination of the weather matrix.

No doubt the time of the aran is past. Few researchers would expect to find a "N-ray" type of substance within the weather matrix that is the sole cause of weather-related behaviors or disorders. There may be biorelevant components of the weather matrix not as yet measured, but presumably they will show the characteristics of other physical variables.

However, within an area that suddenly becomes politically or economically important but at the same time is characterized by immense complexity, there will be a general tendency to resurrect the "simple" explanations and magic substances that explain it all. Whatever they are called, arans, hidden variables, or residuals, they impede the less attractive but immutable necessity to carefully and systematically analyze well collected data.

REFERENCES

Adey, W. R. The influences of impressed electrical fields at EEG frequencies on brain and behavior. In Behavior and brain electrical activity, edited by W. Burch and H. L. Altshuler, 363-90. New York: Plenum Press, 1975.

Altmann, G. Oxygen and biochemical changes following ELF exposures. In ELF and VLF electromagnetic field effects, edited by M. A. Persinger, 227-42. New York: Plenum Press, 1974.

Bachman, C. H., McDonald, R. D., & Lorenz, C. J. Some effects of air ions on the activity of rats. International Journal of Biometeorology, 1966, 10, 39-46.

Barber, D. R. Apparent solar control of the effective capacity of a 110-V. 170 AH lead-acid storage battery in an 11-year cycle. Nature, 1962, 195, 684-88.

Barchas, J., & Usdin, E. (Eds.). Serotonin and behavior. New York: Academic Press, 1973.

Barnwell, F. H. A day-to-day relationship between oxidative metabolism and world-wide geomagnetic activity. Biological Bulletin, 1960, 119, 303.

Basu, D. Relation between the visibility of Jupiter's Red Spot and solar activity. Nature, 1969, 222, 69-70.

Bauer, S. F., & Hornick, E. J. Lunar effect on mental illness: the relationship of moon phase to psychiatric emergencies. American Journal of Psychiatry, 1968, 125, 696-97.

Becker, R. O. The neural semiconduction control system and its interaction with applied electrical current and magnetic fields. Proc. 11th International Cong. Radiol., 1965, 11, 1,753-59.

Becker, R. O. The effect of magnetic fields upon the central nervous system. In Biological effects of magnetic fields, edited by M. F. Barnothy, 207-14. New York: Plenum Press, 1969.

307

Becker, R. O. Electromagnetic fields and life. Technology Review, 1972 (December), 32-37.

Beynon, W. J. G., & Winstanley, E. H. Geomagnetic disturbance and the troposphere. Nature, 1969, 222, 1,262-63.

Bigg, E. K. Lunar and planetary influences on geomagnetic disturbances. Journal of Geophysical Research, 1963, 68, 4,099-104.

Blackman, S., & Catalina, D. The moon and the emergency room. Perceptual and Motor Skills, 1973, 37, 624-26.

Blanco, C., & Catalano, S. Correlation of the air temperature at Catania and the sunspot cycle. Journal of Atmospheric and Terrestrial Physics, 1975, 37, 185-87.

Boles, R. S., & Westerman, M. P. Seasonal incidence and precipitating causes of hemorrhage from peptic ulcer. Journal of the American Medical Association, 1954, 156, 1,379-83.

Bowen, E. G. A lunar effect on the incoming meteor rate. Journal of Geophysical Research, 1963, 68, 1,401-03.

Boyd, J. T. Climate, air pollution, and mortality. British Journal of Preventive and Social Medicine, 1960, 14, 123-35.

Brezowsky, H., & Ranscht-Froemsdorff, W. R. Myocardial infarcts and atmospherics. Z. angew. Bader.-u. Klimaheik, 1966, 13, 679-86.

Brierley, D. M., & Davies, J. G. Lunar influence on meteor rates. Journal of Geophysical Research, 1963, 68, 6,213-16.

Brodie, B. B., & Shore, P. A. A concept for a role of serotonin and norepinephrine as mediators in the brain. Annals of the New York Academy of Sciences, 1957, 66, 631-51.

Brown, F. A., Jr. How animals respond to magnetism. Discovery, 1963 (November).

Brown, F. A., Jr. Propensity for lunar periodicity in hamsters and its significance for biological clock theories. Society for Experimental Biology and Medicine, 1965, 120, 792-97.

Brown, F. A., Jr., & Park, Y. H. Association-formation between photic and subtle geophysical stimulus patterns—a new biological concept. The Biological Bulletin, 1967, 132, 311-19.

Brown, R. New worries about unheard sound. New Scientist, 1973, 60 (87), 414-16.

Bull, G. M. Meteorological correlates with myocardial and cerebral infarction and respiratory disease. British Journal of Preventive and Social Medicine, 1973, 27, 108-13.

Bureau, R. A., & Craine, L. B. Sunspots and planetary orbits. Nature, 1970, 228, 984.

Burr, H. S. Blueprint for immortality: The electric patterns of life. London: Neville Spearman, 1972.

Cassell, E. J., Lebowitz, M. D., Mountain, I. M., Lee, H. T., Thompson, D. J., Wolter, D. W., & McCarroll, J. R. Air pollution, weather, and illness in a New York population. Archives of Environmental Health, 1969, 18, 523-30.

Cerbus, G., & Dallara, R. F., Jr. Seasonal differences of depression in mental hospital admissions as measured by the MMPI. Psychological Reports, 1975, 36, 737-38.

Cook, R. K. Strange sounds in the atmosphere. Sound, 1962, 1, 12-16.

Dalen, P. Season of birth. Amsterdam: North Holland Publishing, 1975.

Davitz, J. R. The language of emotion. New York: Academic Press, 1969.

Delgado, J. M. R. Modulation of emotions by cerebral radio stimulation. In Physiological correlates of emotion, edited by P. Black, 189-204. New York: Academic Press, 1970.

De Lorge, J., & Marr, M. J. Operant methods assessing the effects of ELF electromagnetic fields. In ELF and VLF electromagnetic field effects, edited by M. A. Persinger, 145-76. New York: Plenum Press, 1974.

Dewey, E. R. Cycles, the science—Part III: The moon as a cause of cycles. Cycles, 1959, 10, 197-205.

Dewey, E. R. Cycles: Selected readings. Pittsburgh: Foundation for the Study of Cycles, 1970.

Dolezalek, H. Some hints for experiments concerning atmospheric electric systems on biological systems, 1-26. Alexandria, Virginia: Subcommission VII of the International Commission on Atmospheric Electricity, 1977.

Donle, W. Dependence of outbreaks of influenza on season and weather. Infection, 1975, 3, 23-27.

Dowse, H. B., & Palmer, J. D. Entrainment of circadian activity rhythms in mice by electrostatic fields. Nature, 1969, 222, 564-66.

Ehrmann, V. W., Leitner, H. V., Ludwig, W., Persinger, M. A., Sodtke, W., & Thomas, R. Therapy with ELF-magnetic fields. Physikalische Medizin, 1976, 5, 161-70.

Ellis, F. P. Mortality from heat illness and heat-aggravated illness in the United States. Environmental Research, 1972, 5, 1-58.

Faust, V. V. Suicide and weather. Fortschritte der Medizin, 1976, 94, 363-69.

Faust, V., & Sarreither, P. Season and psychiatric disorders. Medizinische Klinik, 1975, 70, 467-73.

Faust, V. V., Weidmann, M., & Wehner, W. The influence of meteorological factors on children and youths: A 10 percent random selection of 16,000 pupils and apprentices of Basle City, Switzerland. Acta Paedopsychiatrica, 1974, 44, 150-56.

Folk, G. E. Textbook of environmental physiology. Philadelphia: Lea and Febiger, 1974.

Fraser-Smith, A. C. A possible full-moon enhancement of hydromagnetic emission activity. Journal of Geophysical Research, Space Physics, 1969, 74, 2,987-95.

Fraser-Smith, A. C. Solar cycle control in the 27-day variation of geomagnetic activity. Journal of Geophysical Research, 1973, 78, 5,825-29.

Friedman, H., Becker, R. O., & Bachman, C. H. Geophysical parameters and psychiatric hospital admissions. Nature, 1963, 200, 626-28.

Friedman, H., Becker, R. O., & Bachman, C. H. Effect of magnetic fields on reaction time performance, Nature, 1967, 213, 949-56.

Friedman, H., & Taub, H. A. The transcephalic d.c. potential and reaction time performance. Psychophysiology, 1969, 5, 504-09.

Gauvreau, V. Infrasound. Science Journal, 1968, 4, 33-37.

Gauvreau, V., Condat, R., & Saul, H. Infrasound: generators, detectors, physical properties and biological effects. Acustica, 1966, 17, 1-10.

Geller, S. H., & Shannon, H. W. The Moon, weather, and mental hospital contacts: Confirmation and explanation of the transylvania effect. (Toronto: Lakeshore Psychiatric Hospital.) Unpublished manuscript, 1973.

Gellhorn, E., & Loofbourrow, G. Emotions and emotional disorders: A neurophysiological approach. New York: Harper & Row, 1963.

Goerke, V. H., & Woodward, M. W. Infrasonic observation of a severe weather system. Monthly Weather Review, 1966, 94, 395-98.

Goldstein, K. M. Weather, mood, and internal-external control. Perceptual and Motor Skills, 1972, 35, 786.

Green, G. H. The effect of indoor relative humidity on absenteeism and colds in schools, unpublished report, University of Saskatchewan. 1966.

Green, J. E., & Dunn, F. Correlation of naturally occurring infrasonics and selected human behavior. Journal of the Acoustical Society of America. 1968, 44, 1,456-57.

Guyton, A. C. Textbook of medical physiology. Toronto: W. B. Saunders, 1971.

Hansen, J. B. The relations between barometric pressure and the incidence of peripheral embolism. International Journal of Biometeorology, 1970, 14, 391-97.

Hansen, J. B., & Pedersen, S. A. The relationship between barometric pressure and the incidence of perforated duodenal ulcer. International Journal of Biometeorology, 1972, 16, 85-91.

Hayward, J. N. The amygdaloid nuclear complex and mechanisms of release of vasopressin from the neurohypophysis. In The neurobiology of the amygdala, edited by B. E. Eleftheriou, 685-739. New York: Plenum Press, 1972.

Hefferline, R. F., Bruno, L. J. J., & Camp, J. A. Hallucinations: An experimental approach. In The psychophysiology of thinking, edited by F. J. McGuigan and R. A. Schoonover, 299-342. New York: Academic Press, 1973.

Hefferline, R. F., & Perera, T. B. Proprioceptive discrimination of a covert operant without its observation by the subject. Science, 1963, 139, 834-35.

Hensel, H. Neural processes in thermoregulation. Physiological Review, 1973, 53, 948-1, 017.

Herin, R. A. Electroanesthesia: A review of the literature (1819-1965). Activitas nervosa superior, 1968, 10, 439-54.

Heyer, H. E., Teng, H. C., & Barris, W. The increased frequency of acute myocardial infarction during summer months in a warm climate: A study of 1,386 cases from Dallas, Texas. American Heart Journal, 1953.

Hollander, J. L. Environment and musculoskeletal diseases. Archives of Environmental Health, 1963, 6, 527-36.

Huntington, E. Mainsprings of civilization. New York: John Wiley, 1959.

Israel, H. Atmospheric electricity: Volume II—fields, charges, currents. Jerusalem: Israel Program for Scientific Publications, 1973.

Jacobs, J. A., & Atkinson, G. Planetary modulation of geomagnetic activity. In Magnetism and the cosmos, edited by W. R. Hindmarsh, 402-14. London: Oliver and Boyd, 1967.

Katayama, K., & Momiyama-Sakamoto, M. A biometeorological study of mortality from stroke and heart diseases: Its geographical differences in the United States. Meteorology and Geophysics, 1970, 21, 127-39.

Knoll, N., Eichmeier, J., & Shon, R. W. Properties, measurements, and bioclimatic action of small multimolecular atmospheric ions. Advances in Electricity and Electronic Physics, 1964, 19, 177-254.

Koch, W. Humidity sensations in the thermal comfort range. Architectural Science Review, 1963 (March), 33-34.

König, H. L. Concerning the influence on the environment of ELF occurrences in the atmosphere. Z. angew Bader.-u. Klimaheilk, 1962, 9, 481-501.

König, H. L. Behavioral changes in human subjects associated with ELF electric fields. In ELF and VLF electromagnetic field effects, edited by M. A. Persinger, 81-100. New York: Plenum Press, 1974.

Krueger, A. P., Andriese, P. C., & Kotaka, S. Small air ions: Their effect on blood levels of serotonin in terms of modern physical theory. International Journal of Biometeorology, 1968, 12, 225-39.

Krueger, A. P., & Kotaka, S. The effects of air ions on brain levels of serotonin in mice. International Journal of Biometeorology, 1969, 13, 25-38.

Krueger, A. P., Kotaka, S., & Reed, E. J. The course of experimental influenza in mice maintained in high concentrations of small negative air ions. International Journal of Biometeorology, 1971, 15, 5-10.

Krueger, A. P., & Reed, E. J. Biological impact of small air ions. Science, 1976, 193, 1,209-13.

Lebowitz, M. D. A comparative analysis of the stimulus-response relationship between mortality and air pollution-weather. Environmental Research, 1973, 6, 106-18. (a)

Lebowitz, M. D. A comparison of the relationships of mortality and morbidity with air pollution-weather and the implications for further research. The Science of the Total Environment, 1973, 2, 191-95. (b)

Lester, W., Jr. The influence of relative humidity on the infectivity of air-borne influenza A virus (PR8 strain). Journal of Experimental Medicine, 1948, 88, 361-69.

Lethbridge, M. D. V. Relationship between thunderstorm frequency and lunar phase and declination. Journal of Geophysic Research, 1970, 75, 5,149-54.

Lieber, A. L. Human aggression and the lunar synodic cycle. Journal of Clinical Psychiatry, 1978, 39, 385-87. (a)

Lieber, A. L. The lunar effect: Biological tides and human emotions. Garden City, N.Y.: Anchor Press, 1978. (b)

Lieber, A. L., & Sherin, C. R. Homicides and the lunar cycle: Toward a theory of lunar influence on human emotional disturbance. American Journal of Psychiatry, 1972, 129, 101-06.

Lipa, B. J., Sturrock, P. A., & Rogot, E. Search for correlation between geomagnetic disturbances and mortality. Nature, 1976, 259, 302-04.

Lotmar, R., & Ranscht-Froemsdorff, W. R. Intensity of tissue respiration and weather factors (correlation of QO_2 of rabbitskin and atmospherics). Z. angew Bader.-u. Klimaheilk, 1968, 15, 1-10.

Lotmar, R., Ranscht-Froemsdorff, W. R., & Weise, H. Damping of respiration of mouse liver tissue by artificial impulsed radiation. International Journal of Biometeorology, 1969, 13, 231-38.

Ludwig, H. W. A hypothesis concerning the absorption mechanism of atmospherics in the nervous system. International Journal of Biometeorology, 1968, 12, 93-98.

Ludwig, H. W. The influence of ELF electromagnetic fields upon higher organisms. Biomedical Technology, 1971, 16, 67-72.

Ludwig, H. W. Effect of a screening of the electric field during the night on rheumatism sufferers. Archives Meteorologie Geophysics Bioklimatologie, 1973, 41, 305-11.

Ludwig, H. W. Electric and magnetic field strengths in the open and in shielded rooms in the ULF- to LF-zone. In ELF and VLF electromagnetic field effects, edited by M. A. Persinger, 35-80. New York: Plenum Press, 1974.

MacLean, P. D. The limbic brain in relation to the psychoses. In Physiological correlates of emotion, edited by P. Black, 129-46. New York: Academic Press, 1970.

Maczynski, B., Tyczka, S., Marecki, B., & Gora, T. Effect of the presence of man on the air ion density of an office room. International Journal of Biometeorology, 1971, 15, 11-21.

Malek, J., Gleich, J., & Maly, V. Characteristics of the daily rhythm of menstruation and labor. Annals of the New York Academy of Science, 1962, 96, 1,042-55.

McIntyre, D. A., & Griffiths, I. D. Subjective responses to atmospheric humidity. Environmental Research, 1975, 9, 66-75.

Menaker, W., & Menaker, A. Lunar periodicity in human reproduction: A likely unit of biological time. American Journal of Obstetrics and Gynecology, 1959, 77, 905-14.

Nelson, J. H. Shortwave radio propagation correlation with planetary positions. RCA Review, 1951, 12, 26-34.

Neutra, R. Meteorological factors and eclampsia. Journal of Obstetrics and Gynecology of the British Commonwealth, 1974, 81, 833-40.

Novikova, K. F., Gnevyshev, M. N., Tokareva, N. V., Ol, A. I., & Panov, T. N. The effect of solar activity on the development of myocardial infarction and mortality resulting therefrom. Cardiology, 1968, 4, 109-12.

Ormenyi, I. Possible effects of ELF range atmospherics of 3-cps range on traffic accidents in a metropolitan area in Hungary. International Journal of Biometeorology, 1972, 16 (supplement), 93-94.

Ossenkopp, K. P., & Ossenkopp, M. D. Self-inflicted injuries and the lunar cycle: A preliminary report. Journal of Interdisciplinary Cycle Research, 1978, 4, 337-48.

Persinger, M. A. ELF and VLF electric and magnetic field effects: The patterns and the problems. In ELF and VLF electromagnetic field effects, edited by M. A. Persinger, 275-310. New York: Plenum Press, 1974.

Persinger, M. A. Day time wheel running activity in laboratory rats following geomagnetic event of 5-6 July, 1974. International Journal of Biometeorology, 1976, 20, 19-22.

Persinger, M. A., Carrey, N. C., Lafrenière, G. F., & Mazzuchin, A. Thirty-eight blood, tissue, and consumptive measures from rats exposed perinatally and as adults to 0.5 Hz magnetic fields. International Journal of Biometeorology, 1978 22, 213-26.

Persinger, M. A., Lafrenière, G. F., & Ossenkopp, K. P. Behavioral, physiological, and histological changes in rats exposed during various developmental states to ELF magnetic fields. In ELF and VLF electromagnetic field effects, edited by M. A. Persinger, 177-226. New York: Plenum Press, 1974.

Persinger, M. A., Ludwig, H. W., & Ossenkopp, K. P. Psychophysiological effects of extremely low frequency electromagnetic fields: A review. Perceptual and Motor Skills, 1973, 36, 1, 131-59.

Persinger, M. A., Lafrenière, G. F., and Mainprize, D. N. Human reaction time variability changes from low intensity 3-Hz and 10-Hz electric fields: interactions with stimulus pattern, sex, and field intensity. International Journal of Biometeorology, 1975, 19, 56-64.

Pittock, A. B. A critical look at long-term sun-weather relationships. Reviews of Geophysics and Space Physics, 1978, 16, 400-20.

Ranscht-Froemsdorff, W. R., & Rinck, O. Electromagnetic manifestations of the Foehm (correlation between blood clotting and artificial sferics programs). Z. angew. Bader-u. Klimaheilk, 1972, 19, 169-76.

Ravitz, L. J. Electrodynamic field theory in psychiatry. Southern Medical Journal, 1953, 46, 650-60.

Ravitz, L. J. History, measurement, and application of periodic changes in the electromagnetic field in health and disease. Annals of the New York Academy of Science, 1962, 98, 1,141-1,201.

Rechtschaffen, A., Lovell, R. A., Freedman, D. X., Whitehead, W. E., & Aldrich, M. The effect of parachlorophenylalanine on sleep in the rat: Some implications for the serotonin sleep hypothesis. In Serotonin and behavior, edited by J. Barchas and E. Usdin, 401-18. New York: Academic Press, 1973.

Reiter, R. Atmospheric electricity and natural radioactivity. In Medical climatology, edited by S. H. Licht, 280-316. New York: Waverly, 1964.

Reiter, R. Increased frequency of stratospheric injections into the tropopause as triggered by solar events. Journal of Atmospheric and Terrestrial Physics, 1976, 38, 503-10.

Richards, J. H., & Marriott, C. Effect of relative humidity on the rheologic properties of bronchial mucus. American Review of Respiratory Disease, 1974, 109, 484-86.

Rim, Y. Psychological test performance of different personality types on sharav days in artificial air ionization. International Journal of Biometeorology, 1977, 21, 337-40.

Roberts, W. O., & Olson, R. H. Geomagnetic storms and wintertime 300-mb trough development in the North Pacific-North America Area. Journal of the Atmospheric Sciences, 1973, 30, 135-40.

Rogers, W., & Randall, W. Multiphasic variations in sunshine and thyroid activity during a year. International Journal of Biometeorology, 1972, 16, 53-69.

Russell, C. T., & McPherron, R. L. Semiannual variation of geomagnetic activity. Journal of Geophysical Research, 1973, 78, 92-108.

Sakamoto, M. M., & Katayama, K. Statistical analysis of seasonal variation in mortality. Journal of the Meteorological Society of Japan, 1971, 49, 494-509.

Schacter, S., & Singer, J. Cognitive, social, and physiological determinants of emotional state. Psychological Review, 1962, 69, 379-99.

Shah, G. M. Lunar influence on atmospheric ozone. Nature, 1972, 237, 275.

Silverman, D., & Kornblueh, I. H. The effect of artificial ionization on the air on the electroencephalogram. American Journal of Physical Medicine, 1957, 36, 352-58.

Smith, E. V. P., & Gottlieb, D. M. Solar flux and its variations. Space Science Reviews, 1974, 16, 771-802.

Stephens, R. W. B. Infrasonics. Ultrasonics, 1969 (January), 30-35.

Sulman, F. G., Levy, D., Lunkan, L., Pfeifer, Y., & Tal, E. Absence of harmful effects of protracted negative air ionization. International Journal of Biometeorology, 1978, 22, 53-58.

Takahashi, K. The effect of the change of solar activity to meteorological phenomena. Journal of the Meteorological Society of Japan, 1958, 36, 97-107.

Takahashi, K. Key day analysis of the relationship between solar activity and precipitation. Journal of the Meteorological Society of Japan, 1966, 44, 246-54.

Takahashi, K. A short supplementary note on the relationship between solar activity and daily amount of precipitation. Journal of the Meteorological Society of Japan, 1968, 46, 152.

Taylor, L. J., & Diespecker, D. D. Moon phases and suicide attempts in Australia. Australian Psychological Reports, 1972, 31, 110-20.

Teng, H. C., & Heyer, H. E. The relationship between sudden changes in weather and the occurrence of acute myocardial infarction. American Heart Journal, 1955, 49, 9-20.

Thom, E. C. The discomfort index. Weatherwise, 1959, 12 (2), 57-60.

Traute, V., & Düll, B. Correlation between geomagnetic storms and death rate. Deutsche Medizinische Wochenschrift, 1935 (January), 95.

Tromp, S. W. Medical biometerology, New York: Elsevier, 1963.

Tromp, S. W. Seasonal and yearly fluctuations in meteorologically induced electromagnetic wave patterns in the atmosphere (1956-1968) and their possible biological significance. Journal of Interdisciplinary Cycle Research, 1970, 1, 193-99.

Tromp, S. W. The relationship of weather and climate to health and disease. In Environmental medicine, edited by G. M. Howe and J. A. Loraine, 72-79. London: W. Heinemann, 1973.

Tromp, S. W., & Bouma, J. Possible relationship between weather, hexosamine excretion, and arthritic pains. International Journal of Biometeorology, 1966, 10, 105-06.

Unger, M. Hay fever, humidity, and sinusitis. New York State Journal of Medicine, 1950, 50, 1,727-30.

Unger, M. Hay fever, humidity, and sinusitis: Further studies. New York State Journal of Medicine, 1953, 53, 1,674-76.

Weiskott, G. N. Moon phases and telephone counselling calls. Psychological Reports, 1974, 35, 752-54.

Weatherwise, 1971 (June).

Wever, R. ELF-effects on human circadian rhythms. In ELF and VLF electromagnetic field effects, edited by M. A. Persinger, 101-44. New York: Plenum Press, 1974.

Wilcox, J. M. Solar activity and the weather. Journal of Atmospheric and Terrestrial Physics, 1975, 37, 237-56.

Wilcox, J. M. Solar structure and terrestrial weather. Science, 1976, 192, 745-48.

Williams, D. H., & Cohen, E. Human thresholds for perceiving sudden changes in atmospheric pressure. Perceptual and Motor Skills, 1972, 35, 437-38.

Willis, D. M. The energetics of sun-weather relationships: Magnetospheric processes. Journal of Atmospheric and Terrestrial Physics, 1976, 38, 685-98.

Wood, K. D. Sunspots and planets. Nature, 1972, 240, 91-92.

Woodworth, R. S., & Schlosberg, H. Experimental psychology. New York: Holt, Rinehart & Winston, 1962.

INDEX

A-indexes, 229
absenteeism and infrasound, 203
action potentials, 19 [?]
Adey, W. R., 259
age and weather effects, 149, 295-96
aggression and the limbic system, 176-79
air mass, 2; and human complaints, 100-2
amorphous theory, 304-6 (see also theory)
amygdala, 177-79, 223
anticipation and the frontal cortex, 179-80
anxiety, 68, 75, 179-80
apprehension: conditioning of, 67-68
aran, 305
arthritis: and the climatron, 189-91; and ELF fields, 189-91
astrogeophysical correlates of weather, 15-16
atmospheric ions, 213-26; pressure, 182-93; phenomena and the moon, 275-76
autonomic behaviors, 23-26, 303-4
averaging, problems, 120

barometric pressure, 8-29; and bioconsequences, 182-93; and planetary position, 289; and solar disturbances, 268-69
behaviors: correlated with weather, 3, 99-100, 294-95; optimal measurements, 298-300
bioelectromagnetism, 34

biological sites of weather action, 17-34
biomagnetic effects, 235-57
birth: and ambient fields, 255; and lunar phase, 280
blood, 26-27, 295; and electric fields, 256; and magnetic fields, 249-50, 254; and temperature, 173-76
bones, 27-28
brain, 18-23
breezes, 194-95
brightness: bioconsequences, 171-72
Brown, Frank, studies, 277-78

C-index, 229
cardiovascular system, 26
causality, 111
causation versus correlation, 109-10
cell membrane, 30-31
central nervous system, 18-23
cerebrum and limbic system interface, 179-80
charges: electric, 213-26
Chinook, 198
chondroitin sulfates, 31, 259
cingulate gyrus, 177-79
circadian clocks, 57-58; and ELF magnetic fields, 250-51
classical conditioning and biometeorology, 76-97
climatron, 189-91
climate and solar activity, 270
cloudy days: bioconsequences, 170
cluster effects, 5, 303-4
cognitive aspects of experience, 90-97
cold and warm fronts, 156-57
comfort and humidity, 168-69

common cold and temperature, 157-60
compensation and overshoots, 53-54
conditioning: 20-21, 62-76; as confounding factors, 76-79; of physiological/hematological responses, 67
conduction, 133
confounding factors, 112-13, 149-50
connective tissue, 32
contrast enhancement, 285
convection, 133
correlation: limits, 112-18; method, 102-4; of response systems, 72-75
corticosteroids, 77-78
covert operant conditioning, 75-76
current: induction, 213
curvilinear effects, 116, 118
cussing: as superstitious behavior, 71
cycles: in solar activity, 264-65

d.c. potential and psychiatric behaviors, 283-84
death categories and heat waves, 154-55
depression and serotonin, 222-26
determinants of behavior: words, 89-91
discrimination of details, 89
discrete operations: as words, 85-88; responses in weather, 100
diseases: and humidity, 166-68; and expectancy, 95-97
diurnal variation, 25
dream-like condition and temporal lobe, 178-79
dreams and serotonin, 222-26
durations of exposure, 162-63

eclampsia, 161-62
egocentricity, 93
electric fields, 11-12, 33-34, 207-13; bioconsequences, 209-10; biological sources, 33-34, 211-12
electromagnetic fields, 12-13, 227-60; biosources, 33-34; pollution, 251
embolism and barometric pressure, 186-87
emotions, 21-23, 303-4; and ions, 217-19; and limbic structures, 176-79; and serotonin, 222-26
energy, 129-30; from sun, 266-67
eosinophils: conditioning levels, 67
epileptic failure, 41
evaporative cooling and humidity, 166
expectancy, 90-97; in ion studies, 219-21; and weather effects, 96-97
extinction, 65-66
extremely low frequency electromagnetic fields: characteristics, 12, 240-43; bioconsequences, 243-57; emissions, 241
extremely low frequency sound, 199-206
extremes of temperature, 151-56

Faraday cage exposures, 256
feedback models: application to weather, 35-61
flares, 265-66; and weather, 272-73
Foehn, 197-98
frontal cortex and weather effects, 179-81
fronts, 3; bioconsequences, 156-57

gases from soil, 13
geomagnetic disturbances: and Mercury and Venus, 288-89; and weather, 233-34
geomagnetic storms, 15, 202, 229-32, 276

geomagnetism: biorelevance, 235
geophysical effects, 14-15
geriatric populations and weather, 295-96
ground substance, 31-32

hallucinations, 39
headaches and serotonin, 222-26
health and ELF fields, 250-51
heart failure, 40, 139-46, 244
heat waves, 152-56
Hefferline data, 75-76
hematocrit and ELF electric fields, 256
hemoglobin, 258
heparin, 31, 259
hidden variable, 110
hippocampus, 176-79, 258; and serotonin, 223
Hollander experiments, 189-91
homeostatic theory, 35-43, 173-76
homeostats: biological example, 37-38; intrinsic properties, 40-42; mechanical example, 37; psychological example, 38-39
homicides and lunar cycles, 279-80
human behavior and geomagnetic storms, 235-36; and lunar cycles, 278-80
human being, description, 17
humidity, 78; and comfort, 168-69; and electrostatics, 164; and lunar effect, 284
humidity effects, 163-69
hunger, 38; and hypothalamus, 21-23
hyaluronic acid, 31, 259
hydrogen, ionized, 263-64
hydromagnetic emissions, 275
hypothalamus, 6-7, 21-23, 173-76
hypoxia, 257-58

immune system, 294
impulse signals, 47-48
infancy, 65
influenza and temperature, 157-60
infrasound: bioconsequences, 204-5; characteristics, 16, 199-205; sources, 201-2
inhibition: and cerebrum, 179-81; and ELF magnetic fields, 253
interactions, 5
invalids and weather effects, 297
invertebrates and lunar cycle, 277
involuntary conditioning, 72-76
ions: in atmosphere, 10-11, 213-26; (atmospheric) and climatron, 189-91; biorelevance, 215-16; and humidity, 217; and sources, 215
irritability: and geomagnetic storms, 235-36; and ions, 216

Jupiter and solar cycle, 272, 286-87

K-indexes, 229
Krueger studies, 221-22

labels and experience, 85-88, 296-97
lag correlations, 115-16
language and measurement, 62-97
law of initial values, 50-53
learning, 62-76; and brain, 20-21; and magnetic field exposure, 251-54; and set points, 42-43; superstitious behavior, 70-72; unconscious, 72-76
Lebowitz model, 59-61
Levanter, 198
Lieber lunar studies, 279
limbic structures, 21; and atmospheric ions, 222-26
limbic system and temperature effects, 176-79
liquid crystals, 31-32, 260

locus ceruleus, 226, 282-83
Lombarde, 198
low gradient systems, 28-34
low-level effects: from weather, 295-96
lunar cycles: bioconsequence, 274-85
lunar phases, 15-16
lunar theory, 282-83
lunar tides and barometric pressure, 186

magnetic fields: interaction with lunar cycle, 277; storms, 229-32
magnetism and concept, 245-47; as pseudotheory, 280
magnetosphere, 228-29
man-made sources of infrasound, 202-3
mast cells, 32
matrix, 4
measurement, language as, 84-97
mechanical stimuli, 182-206
mechanisms (see theory)
median raphe and serotonin, 223
membrane of cell and magnetic fields, 258-59
memory, 83; and magnetic field exposure, 252-54
Mercury and geomagnetic disturbances, 288-89
metaphors as measurements, 88-89
methodology and lunar cycle, 278-79; in biometerology, 99-119; optimal response measurements, 298-99
microseisms, 14
microvibrations, 33
midi, 198
models (see theory)
models for weather effects, 35-43
monochromatic sunspot phenomena, 264

moon, 273-85
morphine, 66
mortality, 161; and geomagnetic storms, 236; and heat waves, 152-56; and temperature, 139-51
multiple regression, 305
myocardial infarct and season, 144-49

natural signals, 48-50
negative feedback, 40; ions and serotonin, 221-223; reinforcement, 70; stimuli, 66-67
Nelson report, 287-88
neurons, characteristics, 19-20
neutral sheet, 229
Noradrenalin, 23

operant procedures, 69-70
organic materials, 13-14
oxidative mechanisms of magnetic field effects, 257-58
oxytocin, 21, 282
ozone and lunar cycles, 275

parachlorophenalalanine (PCPA), 223-24
paramagnetism and biosystems, 257-58
parasympathetic effects, 23-26
paraventricular nucleus of hypothalamus, 282
Pearson product-moment correlation, 104-9
people and sensitivity, 118-23
perinatal exposure to magnetic fields, 255
periodicities, in geomagnetic variation, 232-33
phases: of moon, 274; of weather, 123-28
photosphere, 262
physiology and geomagnetism, 240
placebo and disease, 95-97; effects, 91-97

planetary correlates of weather, 285-90; position, NTH order biocorrelates, 289-90
polarity (reversals), 209
pollutants, 14
pollution: electromagnetic, 251
positive feedback, 40; ions and serotonin, 221-22
potassium in blood and ELF electric fields, 256
potential differences of bioforms, 211-12
practical versus statistical significance, 104-9
precipitation and magnetic storms, 234
private behaviors (thinking), conditioning of, 86-88
procedures of analyses, 114-18
process operations, as metaphors, 88-89
protein polysaccharides, 31-32, 259
psychiatric behaviors and lunar cycles, 278-80
psychoepidemiology and weather, 300-4

radiation, 133
radical conditioning, as explanation of weather effects, 80-81
radio frequency, ELF pulses, 251
radio reception and planetary position, 287-88
rain: bioconsequences, 169-70
ramp signals, 46-47
rate measures, 115
rates of change and peripheral arterial embolism (PAE), 187; in temperature, 134-37
Ravitz model of lunar-D.C. potential effects, 283
reaction time: and ELF fields, 244-45; and magnetic fields, 247-49

recondite injury, 295-96
reinforcement: definition, 69-70
religious belief: as superstitious behavior, 71-72
replication, 118
resonance and infrasound, 200-1
respiration: and humidity, 164-68; rates and ELF EM fields, 244-45
respondent conditioning, 64-68
response: disruption, 68; exhaustion, 54-56; latency, of feedback systems, 41-42; matrix, 299-300; probabilities, control factors, 50-56; profiles, of systems, 301-3; profiles, to temperature, 157
restlessness: and barometric pressure, 186; and electric fields, 210

scales of measurement, 84-85
Schumann resonances, 241
seasonal clocks, 59; effects, 114
seasons: and temperature, 139-51
sensory stimulation, 38-39
septum, 177-79, 223
serotonin: syndrome, 216-17; theory of weather effects, 222-26
set point, 35-37
sex differences in mortality, 147-48
sexual behavior and serotonin, 222-26
sferics, 241, 256-57
signals: applied to barometric pressure, 192-93; shapes, 44-48; special classes, 48-50
single factor correlation, limits, 113-14
skin and conduction of current, 213
social behavior: as psychoepidemiology, 300-4; determination of mood, 219-21
soil gases, 13
solar: constant, 266-67; disturbances, 16; eleven year cycle,

271-72; terrestrial interaction, 267-68; weather correlations, 268-73; wind, 16, 228
spicules, 263
spleen, 259, 294
stability, 41-42, 157
statistical significance, 104-5
steady: potentials and psychiatric behaviors, 283-84; potentials of human beings, 211-12; state fields, of animals, 33-34
step signals, 44-46
stimuli, 64-65
stratospheric injections, 273
stroke and season, 139-50
sudden commencement: magnetic storm, 232
suggestibility, 94-97; in ion studies, 219-21
suicide: and lunar cycles, 279-80; and serotonin levels, 222-26
sun activity and planetary positions, 286-87; and planets, interaction, 285-87; as star, 262-64; and weather, 261-73
sunspots, 264-65; and earth temperature, 271
superstitious behavior, 70-72; conditioning, 81-82
supraoptic nucleus of hypothalamus, 282
sympathetic nervous system, 23-26
systems: high gradient, 17-23; properties, 73-74; theory, 59-61

tectonic strain and infrasound, 201-2
temperature: behavioral effects, 7; biological effects, 6-7; biosignificance, 131-33; characteristics, 6-7; and climatron,

189-91; control, 42-43; control in biosystems, 173-76; effects, 129-81; and ELF magnetic field interaction, 249-50; and hypothalamus, 22-23; simple physics, 130-31, 133-35; summary effects, 292-93; and sunspots, 271; weekly versus seasonal changes, 134-39
temporal increments, 162-63; of weather, 126-28
testicle weights and magnetic field exposure, 255
theory: of barometric pressure bioeffects, 192-93; dangers, 304-6; of infrasound effects, 205-6; of ion effects, 222-26; of lunar-behavioral effects, 280-85; of lunar-weather effects, 276; of magnetic field effects, 257-60; of temperature control, 174-76; and weather sensitivity, 121-23
thermoloss and wind, 195-97
thinking, 83-97
third factors, 110
thought and development, 86-88
thunderstorms, 270; and barometric pressure, 186; and electric fields, 209; and lunar cycle, 275-76; and solar activity, 269-70
thymus, 294; and magnetic field exposure, 255
thyroid activity: and planetary correlates, 290; cluster in weather effects, 294-95
time and temperature, 134-39; increments, 82-83; lagging, 149; optimal intervals of analyses, 48
traffic accidents and infrasound, 203
triglycerides and magnetic field exposures, 254-55
Tromp weather phase method, 124

turbulence, 194

unconditioned stimuli of weather, 79-80
unconscious conditioning, 72-76
Ungeheuer scheme of weather phase, 124-25

vagus nucleus, 177
variability, 116
variable conditioning, 67
vascular changes and weather, 293-94
vasopressin, 21, 282
Venus and geomagnetic disturbances, 288-89
very low frequency (VLF) electromagnetic fields, 240-43
vibrations, 14; body, 33

water: accumulation in body, 192-93; loss and humidity, 165; molecule, 29-30; vapor, 163-66
weather: biological sites of action, 17-34; effects and expectancy, 95-97; effects, low gradient systems, 28-34; and electric field variations, 208-10; and ELF EM emissions, 242; factors in lunar effects, 284-85; and flares, 272-73; and geomagnetic variation, 233-34; and infrasound, 202; and lunar cycles, 275-76; matrix, components, 3-16; matrix direct consequences, 6-13; matrix indirect consequences, 13-15; possible planetary position correlates, 285-90; and psychoepidemiology, 300-4; related complaints, 99-101; related reports, 120-23; responses: controls, 50-56; sensitive people, 118-23; and solar effects, 261-73; and solar variation, 268-73
weekend effects, 118
wind: as pressure stimulus, 196; biobehavioral effects, 9-10; bioconsequences, 195-97
winds, 194-99
words: and causality, 110-11; as determinants of behavior, 89-91

Zeitgebers, 48-50; and ELF fields, 250-51

ABOUT THE AUTHOR

MICHAEL A. PERSINGER is a Full Professor of Psychology and Head of the Neuroscience Research Group at Laurentian University, Sudbury, Ontario, Canada. Born in 1945, he received his Ph.D. from the University of Manitoba (1971), M.A. from the University of Tennessee (1969) and B.Sc. from the University of Wisconsin, Madison (1967). Dr. Persinger has published more than 50 technical articles in psychobiology, biometeorology, methodology, and neuroscience journals. Progress in Biometeorology (Swets and Zetilinger 1976) and Biometeorological Survey (Heyden 1979) contain chapters summarizing his methodological approach. He has also authored or edited five other books including ELF and VLF Electromagnetic Field Effects (Plenum Press 1974) and (with G. F. Lafrenière) Space Time Transients and Unusual Events (Nelson-Hall 1977). Dr. Persinger teaches an undergraduate course in behavioral biometeorology and is a member of the International Society of Biometeorology.